Technical Approaches to Radio Communications

Troubleshooting and Repair

By

Sheldon A. Chrysler

ISBN: 1-4033-0660-5 (e-book)
ISBN: 1-4033-0661-3 (Paperback)

This book is printed on acid free paper.

1stBooks – rev. 01/31/04

Although this book focuses on C.B. radios and C.B. operators, it is not restricted only to the use of C.B. radios. Basically, all radios are similar in design but differ in frequency and mode of operation.

The author of this publication assumes no responsibility whatsoever for damage to radio equipment, as the information contained in this publication relates to tests, repairs and addition of circuitry to radio equipment.

Acknowledgments

I dedicate this book to my late identical twin brother, Barry B. Chrysler, who dedicated his life to young adults. A scholarship fund has been established in the name of my late twin brother, Barry.

I also dedicate this book to my late parents, Louis and Frieda Chrysler. They gave of themselves unselfishly, shaping and inspiring the lives of my late twin brother and me.

Finally, my last dedication is to the many close friends of mine who stood by me during a very difficult part of my life.

FOUNDATION
BONFILS BLOOD CENTER

I will give a portion of the proceeds from the sale of this book to Bonfils Blood Center Foundation, the fundraising arm of Bonfils Blood Center. The day of the Columbine High School tragedy, I was teaching in a Denver Public School nearby when the call came out for the community to donate blood. I realized then the importance of donating blood. That experience began my involvement with Bonfils Blood Center in Denver.

Bonfils Blood Center is a 501(c)(3) non-profit organization that has been saving lives since it was founded in 1943. It supplies more than 80 percent of the blood needed in Colorado and serves as a safety net for other centers when their inventories run short. Bonfils was the second blood center in North America to receive an ISO certification for its high standards in systems design, testing and monitoring.

During the week following the World Trade Center attack, Bonfils Blood Center drew blood from 6,000 donors – 3,500 more than it typically serves. Its community donor center overflowed with caring and support from Colorado citizens, making it possible for Bonfils to make one of the first deliveries of blood to New York City.

The Emily Griffith Foundation, Inc.

I will also give a portion of the proceeds from the sale of this book to the Emily Griffith Foundation scholarship fund for students at Emily Griffith Opportunity School. While teaching at the school, I realized that at least one-fifth of the students were at or below the poverty level and needed financial assistance to start and complete programs.

The Emily Griffith Foundation, Inc. was established in 1990 as a 501(c)(3) non-profit corporation to provide financial support to Emily Griffith Opportunity School in the areas of student scholarships, innovative projects, capital improvement, instructional equipment and endowment. The Emily Griffith Opportunity School has provided a high quality, post-secondary educational resource to the Metro Denver Community for 87 years and serves approximately 15,000 students annually. Through its 350 technical course offerings, 1.5 million adults have bettered themselves, their lives and the Denver community.

Forward

Questions were asked of me, "How can I get greater range transmitting and receiving?" or "my radio keeps blowing fuses, what is the problem?", et cetera, et cetera.

It seems like the questions were endless and it was this steady flow of questions that encouraged me in writing this book.

The purpose of this book is primarily to address some of the most frequently asked questions and to provide you, the reader, with some technical information on radio communication.

It is an introductory approach to the more simple radio problems and in no way a "cure all" for complex problems that can occur in a radio such as troubleshooting a Phase Locked Loop circuit.

I sincerely hope you benefit from the book. Thank you for your interest and patronage.

Mr. Sheldon Chrysler
Denver, Colorado

Table of Contents

Talk back (sidetone)

Talk-back (also referred to as sidetone) is defined as the ability to hear yourself speak from the C.B.'s speaker, when your microphone is keyed.

Four things are needed to install talk-back into your radio. A carbon resistor, solder, a soldering iron and an Ohm meter.

Follow the step by step instructions along with the illustration figure 1 below to make the installation.

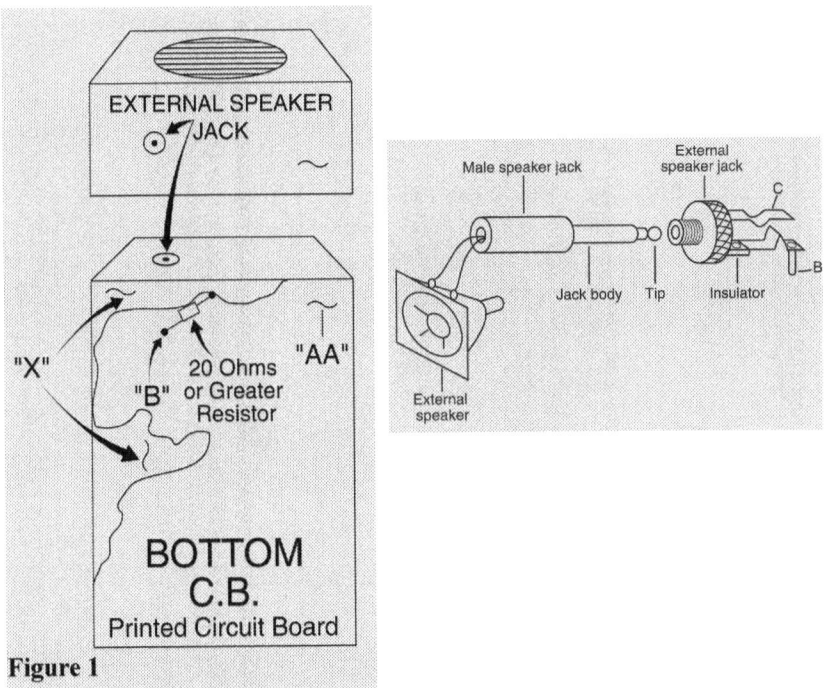

Figure 1

Steps for installing talk-back

1. Remove the C.B. radio from the vehicle.
2. Remove the bottom cover on the C.B. radio to expose the printed circuit board "AA".
3. Locate point "X" on the printed circuit board which is part of the chassis ground. To insure it is ground to the chassis, using an Ohmmeter on the low resistance R x 1 scale, measure from point "C" on the external speaker jack to point "X" on the printed circuit board. You should get a low resistance reading, about one Ohm or less.

4. Solder a 20.0 Ohm, 1/4 Watt carbon resistor to point "B" tip conductor on the printed circuit board, labeled "AA" on Figure 1. Solder the other end of the resistor to the ground marked "X" on Figure 1.
5. Connect the vehicle antenna to the antenna jack on the back of the C.B. radio.
6. Connect the power to the C.B. radio.
7. Turn the C.B. radio on and key up the microphone.
8. Talk or whistle lightly into the microphone. You should hear your voice or whistle over the C.B.'s speaker. If you cannot hear your voice or whistle, check the instructions in 3 through 7.
9. If it is working properly, check to be certain there is enough clearance between the resistor and back cover before replacing the back cover.
10. Disconnect the vehicle antenna.
11. Disconnect the power.
12. Reinstall the C.B. radio in the vehicle.
13. Reconnect the antenna.
14. Reconnect the power.
15. Repeat steps 7 and 8.

Antenna tuning and testing for proper length

Before discussing tuning and testing for proper length of your antenna, let me make a general comment: Metal is a good conductor of electrical energy. Therefore, it stands to reason that a solid metal antenna gives the best results for signal transmitting and receiving. However, because of better conduction, solid metal antennas are noisy.

Fiberglass antennas, on the other hand, are less conductive than solid metal antennas because of less conductive area. There is usually a thin wire running through the center of the fiberglass housing and therefore are not as good for signal transmitting and receiving as solid metal one is because of less conductive contact area on a fiberglass antenna.

With that general comment, let's turn our attention to some of the commonly asked questions about antennas.

1. **Is it okay to place an A.M.-F.M. receiver antenna next to a C.B. antenna?**

A citizen band antenna (or any other transmitting antenna) should be kept as far away as possible from any receiving antenna. Failure to do so will result in receiver damage because of coupled transmitted energy emanating from the transmitting antenna to the receiving antenna.

2. **What is the reason for tuning an antenna to the radio?**

To maximize the power transfer from the radio transmitter and the incoming signal received is the reason for matching an antenna to a radio.

For example, suppose you have an antenna properly tuned to 50 Ohms. If the RMS voltage is 21.0 volts, then power will be
$P = V^2 \div R = (21.0)^2 \div 50 = 8.82$ Watts where P = Power, V = RMS voltage and R = antenna resistance. Now suppose the antenna resistance has changed to 72 Ohms. It then follows that
$P = V^2 \div R = (21.0)^2 \div 72 = 6.1$ Watts. Clearly, less power is delivered to the radiating antenna with the second example. Resistance of an antenna can be determined by measuring the current and voltage at any point on the antenna length at any point in time. By Ohms' law: (Resistance = voltage ÷ current). The resistance of the antenna can be determined: R = radiating resistance of antenna, V = volts RMS and I = current RMS. Voltage across and current into an antenna are measured with special meters designed to operate at radio frequencies.

3

3. Is my antenna too long or too short? What length do I make it? What size and length coaxial cable do I need?

You may be able to determine if your antenna is too long or too short by cupping your hands around the antenna and having someone observe the VSWR (Voltage Standing Wave Ratio, discussed page 75, question 27) meter as the microphone keyed and transmitter is turned on. If the VSWR goes down when transmitting, your antenna is too long and if the VSWR goes up, the antenna is too short. See Figure 2 below:

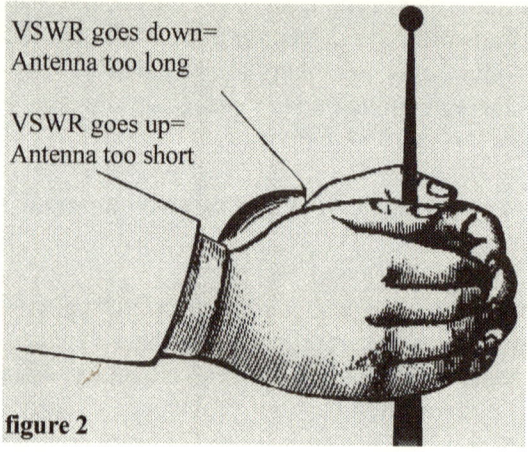

VSWR goes down=
Antenna too long

VSWR goes up=
Antenna too short

figure 2

The antenna length should be only long enough to maintain the standing waves as close to one to one as possible.

Theory of tuning an antenna

An antenna is tuned and will operate at its maximum efficiency and performance when the antenna length is equal to one wave length, multiples or fractions of one wave length of the operating frequency.

For example, suppose there are two tuning forks, TF_1 and TF_2. Each tuning fork will produce 500 vibrations per second when struck with an object. Both tuning forks are set very close to each other. If tuning fork TF_2 is struck and vibrates, it will induce a vibration into the tuning fork TF_1 next to it and vibrate intensity at the rate of 500 vibrations per second.

Now suppose tuning fork TF_2 vibration rate is changed to 100 vibrations per second, tuning fork TF_2 is once again struck and is now vibrating at the new rate of 100 vibrations per second, tuning fork TF_1 vibration rate is still at the initial rate of 500 vibrations per second.

Since there is a difference of vibration rates for tuning forks TF_1 and TF_2, the two different rates will interact destructively and oppose each other's vibrations (forces) and create not as intense vibration as the case when the vibration rates of tuning forks TF_1 and TF_2 are the same.

A similar circumstance occurs when a radio signal is transmitted and received. A radio signal is made up of positive and negative electrical variations that change with respect to time. The changes of these variations is referred to as "frequency".

Similar to the tuning fork example, radio signals must work together in order to receive and transmit a signal at its maximum intensity. An antenna is designed to transmit and receive a signal. It will provide an optimum signal if the antenna has the proper electrical length (refer to question 9b cable length, page 9, section II on antennas. Also see the glossary for the definition of antenna electrical length). To precisely achieve the electrical length of an antenna requires either a wattmeter or a VSWR bridge described in section V, page 45, question 2. Achieving the electrical length of an antenna by either shortening or lengthening it is known as "antenna tuning". Other considerations affecting antenna performance are discussed in section II, page 9, question 9.

4. Why is my receiving and transmitting poor?

Usually, poor receiving and transmitting is the result of a mismatched antenna. This is a very common problem; however, check the voltage standing wave ratio using a standing wave ratio meter. If you get a reading greater than 1.5 to 1, check the connections at the antenna and the antenna jack on the radio. If the connections are good, follow the procedure used for checking whether an antenna is too long or too short (described earlier in this section). A reading of 1.5 or less is a normal acceptable reading. If the reading is not normal, the next step would be to have a radio technician check the problem out.

5. Shall I use one antenna or two?

The use of one antenna is preferable to two for the following reasons:

a. There is less signal insertion loss (that is, losses due to coaxial cable resistance, couplings, et cetera).
b. Concentrated energy is directed to only one antenna.
c. The need to constantly monitor a second antenna for changes in voltage standing wave ratio is eliminated.

6. What is the proper way to tune dual antennas?

When tuning dual antennas, remove one connector from either of the two antennas. Using one barrel connector and one PL-259 connector, solder two carbon 2 watt 100 Ohm resistors in parallel onto the PL-259 connector (see figure 3). Then place this connector with resistors onto the barrel connector and the barrel connector onto the antenna PL-259 cable connector of the disconnected antenna.

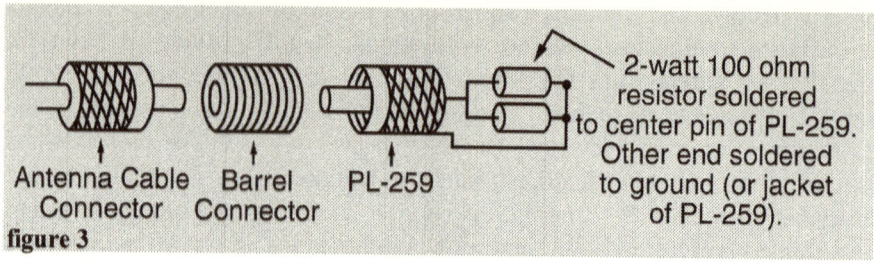

Antenna Cable Barrel PL-259
Connector Connector

2-watt 100 ohm resistor soldered to center pin of PL-259. Other end soldered to ground (or jacket of PL-259).

figure 3

Check the VSWR of this antenna with the above resistive load according to the procedure outlined earlier. Repeat the same process by now placing the resistive load on the other antenna. In the case where no PL-259 is used on the antenna cable but rather an "eye" connector is attached to the antenna cable, the procedure is slightly different. Rather than solder the two carbon 2 watt 100 Ohm resistors onto the PL-259 connectors, solder them onto two alligator clips (see figure 4). Then clip this resistive load onto the "eye" connectors of the antenna cable. Next, repeat the process of checking the VSWR described for the PL-259 connector.

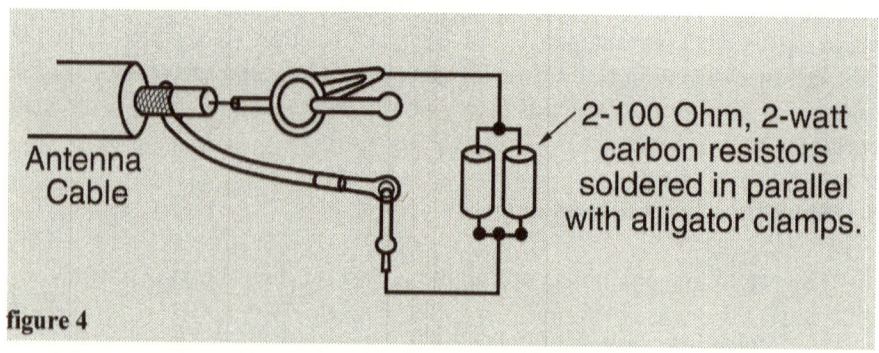

Antenna Cable

2-100 Ohm, 2-watt carbon resistors soldered in parallel with alligator clamps.

figure 4

7. Where is the best place to mount an antenna?

Providing the antenna is placed on a metal surface having a radius approximately equal to the height of the antenna, the best place for the antenna is as near the center of the vehicle cab as possible to adequately provide for equal radiation distribution as illustrated in figure 5 below.

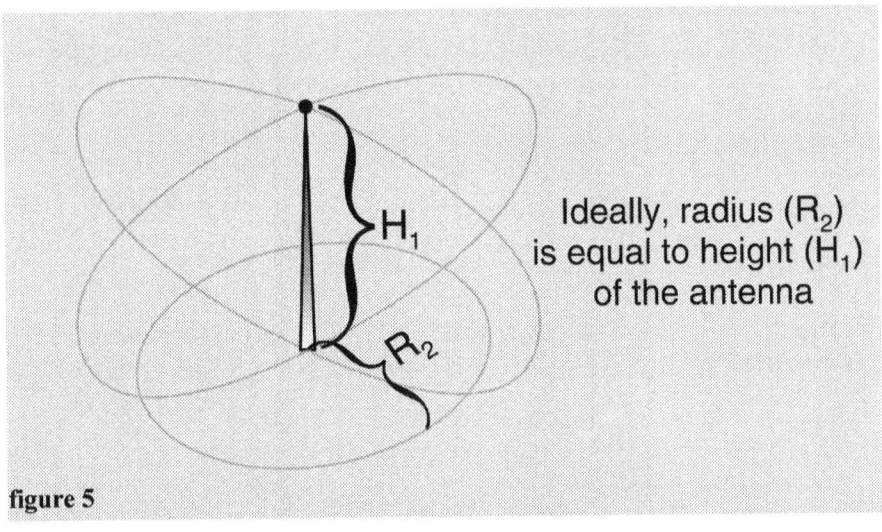

Ideally, radius (R_2) is equal to height (H_1) of the antenna

figure 5

8. **When the cab of a vehicle is fiberglass and has no metallic surface, how is an antenna mounted?**

When there is no metallic surface, antenna gain is substantially reduced and it becomes necessary to provide a ground plane to increase antenna gain (see section II, page 13 question 15). Ground plane refers to any conducting surface (usually horizontal) used in conjunction with an antenna either transmitting or receiving. A ground plane can be made up in several ways. Here are the two most common ways:

The first way of making a ground plane is the use of a metal disk. This disk should be made as large in diameter as possible to maximize the ground plane area. Ideally, the height of the antenna should equal the radius of the disk (see figure 6 on page 8 and also figure 5 above).

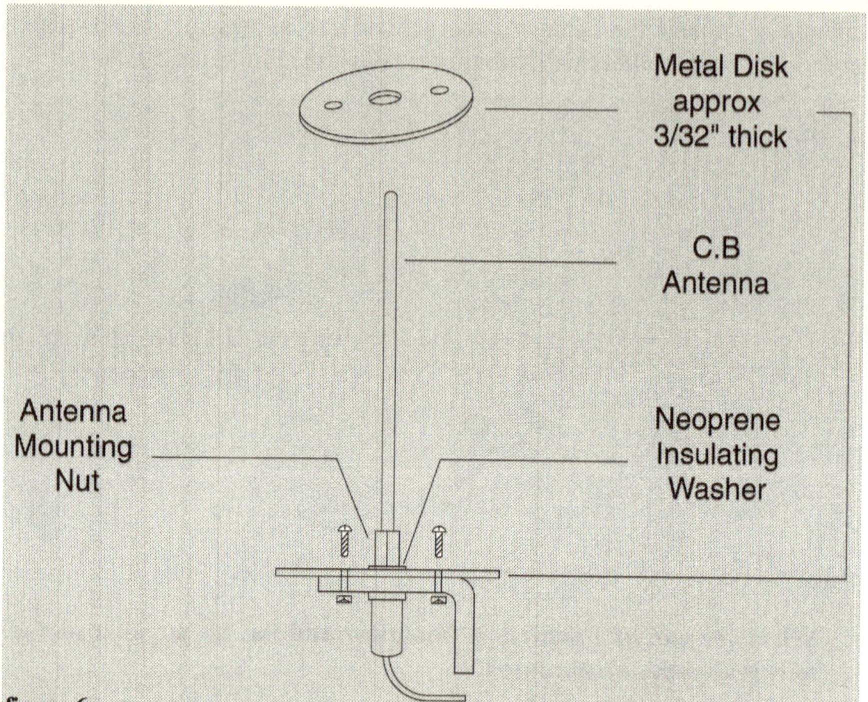

Metal Disk
approx
3/32" thick

C.B
Antenna

Antenna
Mounting
Nut

Neoprene
Insulating
Washer

figure 6

The second way of making a ground plane is using another C.B. antenna. One antenna is grounded, facing downward opposite the vertical upright antenna (see figure 7 below).

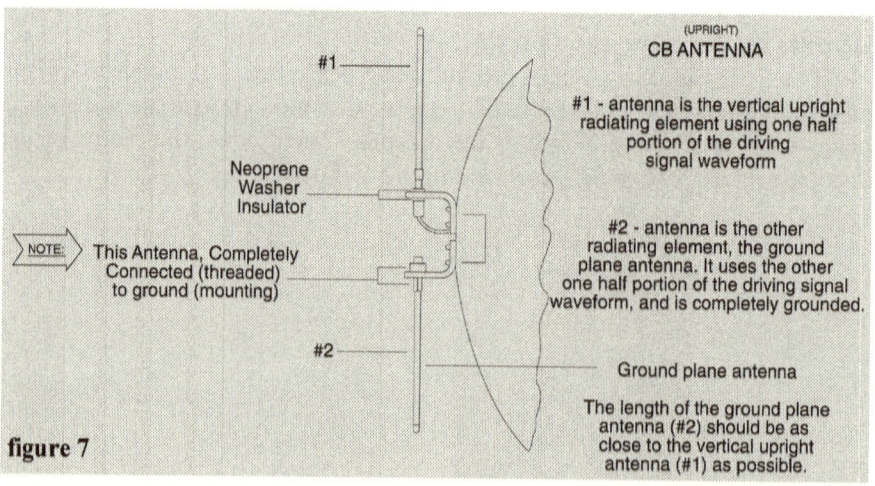

(UPRIGHT)
CB ANTENNA

#1

#1 - antenna is the vertical upright radiating element using one half portion of the driving signal waveform

Neoprene
Washer
Insulator

NOTE: This Antenna, Completely Connected (threaded) to ground (mounting)

#2 - antenna is the other radiating element, the ground plane antenna. It uses the other one half portion of the driving signal waveform, and is completely grounded.

#2

Ground plane antenna

The length of the ground plane antenna (#2) should be as close to the vertical upright antenna (#1) as possible.

figure 7

Neither of these two options, of course, will provide optimum performance of the radio. However, making the metal dish as large as possible in the first option and making the length of two antennas approximately equal in length in the second option, will maximize the antenna gain.

9. Where is the best place to tune my antenna and how can I optimize the performance of my antenna or antennas?

The best place to tune an antenna is in an open area free of buildings and obstructions that would cause reflections of the signal and therefore, giving an erroneous reading on the VSWR meter.

Optimizing the performance of an antenna is a function of having your antenna properly tuned and matched to the output resistance of the radio's transmitter. Here are some helpful hints:

a. Keep the antenna feed line (coaxial cable) as short as possible.

b. Have the cable made up in a length equal to the wave length of the frequency you're operating on. In most cases, Citizen Band frequency is 27 mega hertz (MHz). So, for a full wave length (@ 27 MHz), the full wave equation is: Wave length in feet equals 984 feet divided by frequency in mega hertz. So wave length in feet equals 984 divided by 27, equals 36.4 feet. Since in most cases this would be an impractical length, divide 36.4 feet by 2 which equals 18.2 or a half wave length. Continue to divide by 2 until a practical length is arrived at. In other words, the length could be either 36.4, 18.2, 9.1, 4.55 or 2.275 feet.

Neither bundle up nor coil up the coaxial cable. This can be achieved by the manner in which it is routed.

10. Does tilting my antenna forward help the receiving or transmitting of the radio?

Tilting the antenna does nothing more than reduce the effective radiating power (ERP) by physically reducing the height of the antenna and radiation pattern. An antenna must be in a vertical plane for optimum transmission and reception (see figure 8 on page 10).

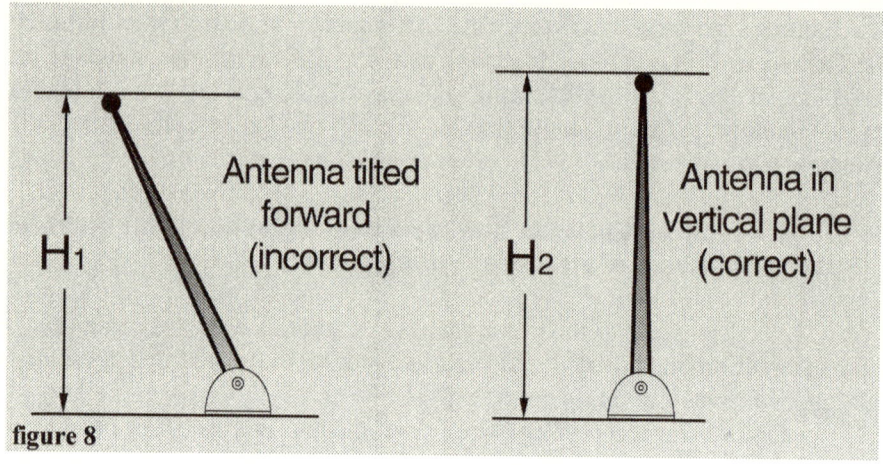

Antenna tilted
forward
(incorrect)
H_1

Antenna in
vertical plane
(correct)
H_2

figure 8

11. May I use a co-phased harness on a single antenna?

A co-phased harness will not work effectively on a single antenna. A co-phased harness is made up of two equally matched cables, one from each of two antennas (coax of the same type and length) tied together at the ground points. They tie to one piece of coaxial cable which connects to the SO-239 connector on the back of the radio. This means then they are tuned for two antennas, not one.

12. My vehicle has a built-in antenna coaxial cable. What kind of performance can I expect of this coaxial cable?

Coaxial cable supplied by the vehicle manufacturer usually is not of very good quality. To ensure optimum performance, C.B. users should elect to purchase a quality C.B. cable/harness from a quality C.B. radio shop or electronics supply company.

13. What is a splitter? Can a splitter be used for an A.M.-F.M. radio from a C.B. radio antenna?

A splitter is an electronic device that matches a radio's signal characteristic from an antenna, then distributes the signal to one or more radio receivers, such as an A.M.-F.M. radio.

While splitters work, a reduction of the radio signal will result when the radio signal is distributed to too many radio receivers. The reduction of a received signal as a result of having too many devices receiving that distributed signal from an antenna is called "loading of the circuit".

A splitter manufacturer will usually indicate the maximum number of radio receivers that can be used from their splitters without significant "loading of the circuit". Keep in mind, however, that an antenna used for transmitting should not be used with a splitter. The transmitted energy going to the antenna will also follow the path into and out of a splitter and end up in the receiving device where the signal is being distributed. It will ultimately damage the device. Therefore, C.B. antennas are not used for receiving a radio signal when that same antenna is being used at the same time to transmit electrical energy. C.B. antennas can be used, however, with a splitter for an A.M.-F.M. receiving antenna as long as it is not being used for transmitting. Remember, only receiving antennas should use a splitter for distribution of a radio signal to be received. The best rule of thumb is to use a separate antenna for each radio. So while splitters do work, they are often used for receiving a radio signal. The splitter matches the electrical characteristics of an antenna and then distributes a matched signal to a respective radio receiver.

14. Why is my antenna warning light on or my VSWR meter reading high?

Either the antenna warning light on or a high VSWR reading on the meter is an indication of usually one of two problems. First, it could be a short and second, it could be a break in the antenna coaxial cable.

To determine if an antenna coaxial cable is shorted, visually inspect both ends of the coaxial cable for breaks, sharp bends and cable being smashed. It is suggested that the coaxial cable be inspected at vulnerable areas (i.e. doors, hoods, vents, etc.) where damage is likely to occur.

In some cases, the terminating end of the coaxial cable going to the antenna is exposed (refer to figure 29 on page 48). In this case, inspect to insure the braided portion of the coaxial cable is not shorted to the center conductor wire or braided wire broken from its ground point on the antenna mounting bracket.

If both ends of the coaxial cable are terminated with a PL-259 connector, one end will go to the antenna, the other to the radio.

It is possible one end of the coaxial cable with the PL-259 connector is shorted within the connector. To make this test, a digital multimeter is needed. If one is available, remove both ends of the connector from the antenna and radio. Using the one times resistance (R x 1) scale connect one lead on the multimeter to the center conductor pin on one of the PL-259 connector and the other lead to the body of the connector. Now, measure the resistance.

Sheldon A. Chrysler

Repeat the above procedure on the other end of the coaxial cable, having a PL-259 connector.

The lowest resistance read on the multimeter from the readings taken on either end of the coaxial cable PL-259 connectors will be the likely connector that is shorted. It is best to cut the PL-259 connector off from the coaxial cable and replace it with a new connector. Cut the connector off about one inch where the cable enters into the connector. Use the procedure outlined in section 8, page 70 to replace the connector.

If a resistance check fails to reveal a shorted cable condition, then an open connection in the coaxial cable is likely.

To determine if the coaxial cable is open, connect one lead of the digital multimeter to the center conductor pin of the PL-259 connector on one end of the coaxial cable and connect the other lead of the multimeter to the other center pin conductor of the PL-259 connector on the other end of the coaxial cable.

If a resistance of several million ohms is read on the digital multimeter, an open connection is suspected. At this point, it is advisable to cut off one of the two PL-259 connectors (about an inch where the cable enters the connector) and using a digital multimeter, on the one times resistance (R x 1) scale measure from the center conductor pin of the uncut PL-259 connector to the center wire of the coaxial cable where the PL-259 connector was cut off.

A resistance reading of about one ohm will indicate the cable is good. If a high resistance reading into the several million of ohms is still read, repeat the same process of cutting off the PL-259 connector on the other end of the coaxial cable, about one inch from where the cable enters the connector. Now measure with the digital multimeter the center wire resistance of each end of the coaxial cable. Again, you should read a resistance of about one ohm. If you still have a resistance reading of several million ohms, then there is a break in the coaxial cable. At this point, discard the coaxial cable and replace the two PL-259 connectors onto the new cable as outlined on replacing connectors in section 8, page 70.

Another possibility (though rare) is the braided wire on the coaxial cable can have an open connection. The same procedure as outlined above for checking the center pin conductors on the PL-259 connectors is used to check for an open connection on the braided (shield) portion on the coaxial cable. The braided portion of the coaxial cable will connect to the PL-259 connector body.

If you ever notice an antenna warning light on or a high VSWR meter reading while keying your transmitter, stop keying the transmitter immediately. If you continue, damage to the transmitter will occur.

15. How can I optimize the radiation pattern of my C.B. radio antenna?

Optimizing the radiation pattern of a C.B. radio antenna requires special considerations.

a. First, you need to strive for the greatest amount of metallic ground plane (see ground plane on page 7, question 8).
b. Make sure your antenna is always tuned to the lowest VSWR. An antenna hitting an obstruction such as a bridge, an underpass, et cetera, will change the antennas electrical characteristics and replacing and retuning will be necessary.
c. Avoid placement of the antenna on a non-conductive surface.
d. Keep antennas as far away as possible from obstacles such as smoke stacks and contours. When mounting an antenna, try to utilize as flat a surface as possible to get a good ground plane.

All of these things need to be considered to optimize the radiation pattern.

16. What does 6 decibel (DB) gain mean that is printed on the box of an antenna?

DB gain can relate to either voltage, current or power. Most generally db gain, when related to an antenna, is concerned with power or voltage. For example, a receiver receives a 250 millivolt signal on a metered indicator from a distant transmitted signal. The receiver will use a reference antenna (call it reference antenna A_1). Another antenna of a different type (call it antenna A_2) using the same receiver now receives a metered signal reading of 1, 000 millivolts (abbreviated 1, 000 mV). Now the gain (or increasing signal) received from the second antenna over the first antenna is said to have a gain of four over that of antenna A_1 to that of which it was referenced. Antenna gain, in terms of voltage, is referred to as "field gain". Here is the mathematical example of field gain:

$$V_G = V_2 \div V_1 \text{ therefore } V_G = 1,000 \div 250 = 4$$
Where: V_G = Voltage Gain
$\quad\quad\quad V_1$ = Smallest signal received (from referenced antenna)
$\quad\quad\quad V_2$ = Largest signal received

Energy emitted from a source in free space contains an internal resistance. This is radiation resistance. Mathematically, it is the voltage divided by the current at any point on the antenna.

Mathematically: Power Gain $(P_G) = P_G = \dfrac{P_2}{P_1}$

P_2 = Power of A_2 antenna
P_1 = Power of A_1 antenna

Also, Power Gain (P_G) can be expressed as

$$P_G = \frac{P_2}{P_1} = \frac{\dfrac{V_2^2}{R}}{\dfrac{V_1^2}{R}} = \frac{V_2^2}{R} \cdot \frac{\cancel{R}}{V_1^2} = \frac{V_2^2}{V_1^2} = \frac{V_2^{\cancel{2}}}{V_1^{\cancel{2}}}$$

Therefore, $P_G = \dfrac{V_2}{V_1}$

Notice the Power Gain formula is the same as the Voltage Gain formula, when the radiation resistance (R) on both antennas A_1, A_2 are close.

You may see antennas rated in terms of D.B. or decibel gain (the explanation of Decibel is beyond the scope of this book; however, its explanation can be found in most radio communication books).

17. How much separation should there be from one transmitting/receiving antenna to another transmitting/receiving antenna?

Ideally, the separation should be one wave length for the frequency being used. Wave-length may be expressed mathematically as follows:

$\lambda = 984 \div F_0$ where F_0 = frequency of the carrier being transmitted in mega hertz. λ = wave-length of the transmitted carrier in "feet".

We use the value 27 MHz for the carrier, since C.B. radios operate on 27 MHz. Therefore, the antenna separation distance must equal 36.5 feet. Since this is an impractical length to use on a vehicle, separate the antennas with as much distance as possible.

14

18. What is the small can on my center loaded antenna?

Inside this can is a coil and the can is filled with oil. The oil is used to maintain fairly constant temperatures of the coil when high power is used. If the coil is allowed to get too hot, changes in the antenna resistance will occur and power output will be erratic and reduced.

1. What's wrong when the radio repeatedly blows fuses?

Inside most radios, a component called a diode is placed on the positive side of the power lead. This diode is to provide for current to flow in one direction depending on the polarity. This diode protects the circuit in the event the positive and negative power leads are accidentally reversed. Therefore, the first thing to check is to see if the diode has a short. This is done by connecting two leads from an Ohm meter to each end of a disconnected diode. Disconnect the diode at one of the points it is soldered onto the terminal board by desoldering (see figure 9 below). If the reading on the Ohm meter is about 1.0 Ohm, the diode is shorted and needs to be replaced. If the reading on the Ohm meter is about 600 Ohms one way, the diode is not shorted. However, one other test must be performed. That is to switch the leads of the Ohm meter to the diode. If the reading on the Ohm meter is about 500, 000 Ohms or greater, the diode is good. If the reading is under 500, 000 Ohms one way, it is advisable to replace the diode.

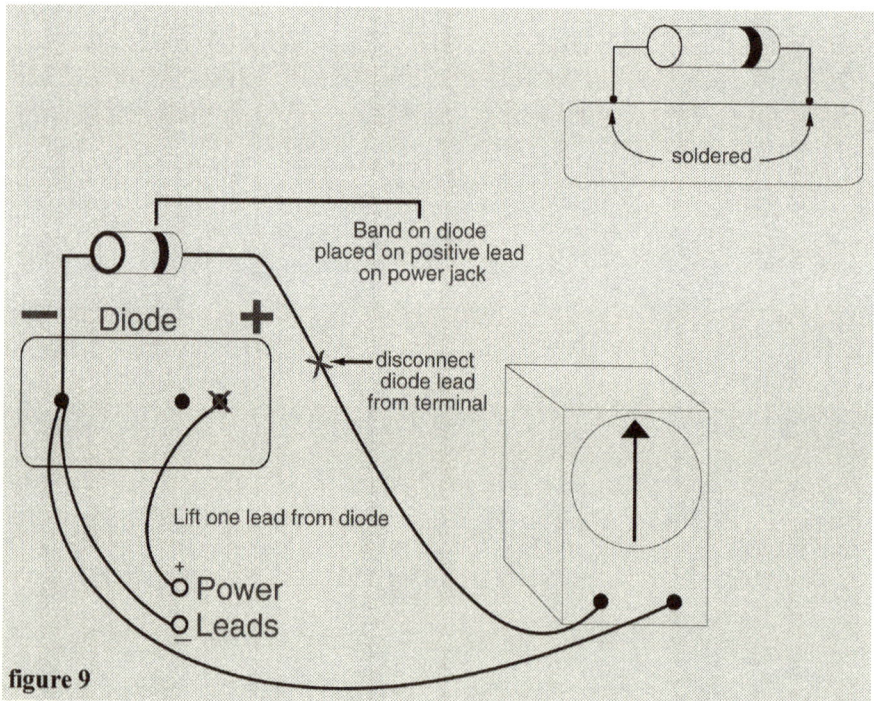

figure 9

Other problems could be occurring in the radio besides the diode conducting; however, this is the most common problem. Some of the other problems, besides the diode being shorted are: a shorted capacitor, a shorted transistor or a chaffed or pinched power lead within the radio chassis. Because of this broad spectrum of possible problems, it is advisable to consult a qualified radio technician for specific diagnostics of the problem.

2. **What is wrong when the receiver functions properly but the transmitter does not?**

   a. First, check to see if the microphone is turned down too far.
   b. Next, check to see if the microphone connector is securely connected.

   If neither of these checks produce favorable results, take the radio to a technician to:

   a. Check whether the power amplifier or the amplifier driver is bad.
   b. Check if the power supply to the amplifier is bad.
   c. Check if the carrier power is too high.
   d. Check for a bad microphone cable.

3. **What is wrong when the transmitter functions properly but the receiver does not?**

   a. First, check to see if the R.F. gain control is turned down too far.
   b. Next, check to see if the squelch is turned up too high.

   If neither of these tests solve the problem, a technician will need to make some of the following tests:

   a. Check for misalignment of the radio receiver.
   b. Check to see if the first radio frequency amplifier is bad in the receiver.
   c. Check to see if there is a bad mixer crystal.

4. **What is wrong when, set on receive, there is no range?**

   With this problem there are five common possibilities:

   a. The R.F. gain is turned down.
   b. The squelch is turned up too high.
   c. The antenna is not tuned properly (see section 1, question 3).

17

Should the above tests not remedy the problem, a technician should be consulted for two other common possibilities:

    a. He should test for a bad receiver alignment.
    b. He should test for a bad radio frequency amplifier in the receiver stage.

5. What is wrong when I talk into the microphone and cannot be heard by the person I am trying to reach?

Some possible reasons for not having range when transmitting are:

    a. Because the output from the transmitter is low. This low output may be the result of a bad final transmitter amplifier or a bad final driver amplifier. Also, it could be wrong voltage going to the transmitter circuit, insufficient radio frequency drive going to either the driver or final transmitter amplifier, or shorted or open component in the transmitter output filter or external coupling to the transmitter (i.e. coaxial cable, antenna, et cetera).

        The above mentioned possibilities for lower power output from a C.B. radio transmitter are complex problems which require the diagnostic ability of an experienced radio technician.

    b. The antenna is not tuned properly (see antenna section).

    c. There is a low microphone audio signal. A low microphone audio signal may be something as simple as the control knob being turned down too far needing only to turn the "mic gain" knob to increase the audio output level. Should you have a novelty device attached to the microphone, remove it and see if this solves the problem, as these kind of devices may reduce the audio output level. It also could be that the microphone has recently been replaced and because every microphone has a different impedance, you may have the wrong microphone. The fourth possibility is to determine if the microphone audio amplifier is defective, which only a qualified radio communications technician can do.

    d. The transmitter is off frequency. If a transmitted signal is different by being off the frequency of what is being received, then the maximum strength of that transmitted signal being received is drastically reduced. To be able to set the C.B. radio on frequency requires the technical skill of an experienced radio technician.

e. A shorted antenna coax. The coaxial cable going to the antenna from the C.B. radio will sometimes get pinched and shorted in the vehicle door, vent or other parts of the vehicle. If the cable is smashed and not shorted, its electrical characteristics are altered and maximum performance of the C.B. radio transmitter is reduced. A shorted coaxial can be determined with an Ohmmeter or continuity tester. Disconnect the PL-259 cable fitting from the radio before testing. When using an Ohmmeter, use the resistance times one scale (R x 1) measuring from the center conductor on the coaxial cable to the braided shield of the cable. If the cable is shorted, you will typically get a reading of about one Ohm or less. If the cable is not shorted and just flattened, a reading of several million Ohms will show on the Ohmmeter. Inspect the coaxial cable from the C.B. radio to the antenna for kinks, breaks and physical damage.

f. The importance of having a properly tuned antenna cannot be over emphasized. Equally important, however, is the need to have the proper length antenna for "capturing" a radio signal to be received and to be transmitted. Ideally, strive to get as long an antenna as possible designed to operate properly on the frequency being transmitted and received. Usually this will provide the most signal gain in transmitting and receiving.

The drawback to a long antenna, however, is of course, its length. There is a practical limit on how long an antenna can be so one must compromise and settle for a practical length. If the reader finds that his antenna is properly tuned, but experiences poor receiving and transmitting, a likely problem would be that the antenna is too short.

6. What is wrong when the radio has carrier but no modulation?

Possible problems why the radio has carrier and no modulation are:

a. The microphone is inoperative because it is miswired. A microphone has what is referred to as a "microphone element" in it. There are two wires that come to the microphone element. Both of these wires come from the microphone cable - one to the ground and one to the speech amplifier. Also, there are about four to six wires that are encased in the microphone cable. These wires control other functions of the radio like the receiver audio. So it is altogether possible to mix these wires up with the wires from the microphone element.

Discussed earlier, one wire of the microphone element will go to the ground and the other wire will go to the speech amplifier. To test if the wires in the microphone circuit have been properly connected, place a PL-259 connector with a number 47 bulb on the back of the C.B. radio and apply power. Then key up the microphone and with a slight whistle into the microphone, note if there is a varying intensity of the number 47 bulb on the back of the radio. If the bulb varies in intensity with the slight whistle, the microphone is connected properly. If no varying intensity is noted with a slight whistle applied, but the bulb continues to illuminate, the microphone is not connected properly.

b. Microphone gain may be turned down too far. On most newer C.B. radios, there is a knob on the front of the face plate marked "mic gain". This controls the level of audio intensity of the microphone. When the knob is turned counter clockwise, usually the level of audio intensity is lower. When the knob is turned clockwise, usually the level of audio intensity is higher. So, if you have a low level of transmitted audio output, check to see if the "mic gain" knob is turned too far counter clockwise.

c. In the case of a power microphone being used, the battery could be bad, the polarity could be reversed or the battery could be missing.

d. A bad audio amplifier in the microphone circuit. When a mechanical sound is changed to an electrical variation in a microphone circuit, this variation is very low in electrical intensity. Therefore, a device is needed to increase or magnify these electrical variations. This device is called an amplifier. An amplifier may be in the form of a transistor, vacuum tube, integrated circuit, transformer, et cetera. If the person being transmitted to receives a carrier, but no audio signal, there is a good likelihood that the sender's amplifier is either bad, poorly connected, or no voltage is being supplied to it. At this point, the circuit needs attention by a qualified radio technician.

e. It could be the wrong microphone. Microphones have different levels of output depending on what type of microphone it is. For example, a dynamic microphone uses a diaphragm with a coil attached to it. This coil is formed over a permanent magnet. When a sound strikes the diaphragm, an electrical variation is induced.

A crystal microphone uses the same principal as the dynamic microphone; however, its electrical variation output is low compared to that of the dynamic microphone. Each microphone is designed for a specific circuit. This is the reason it is important to have the proper microphone for the C.B. radio it was designed for.

f. The transmitter is running too much carrier power. As carrier power increases, audio power must also be increased to maintain the level of modulation proportionally. If the carrier power is increased without increasing the audio power, the level of audio and modulated power are not balanced and the modulation is reduced substantially. (Also refer to section III, page 30, question 15)

7. The radio has no audio and no modulation.

No speaker audio and no microphone transmit audio is sometimes the result of a dirty or bad push-to-talk switch on the microphone or a bad audio power amplifier integrated circuit used in the receiver and transmitter circuit. Figure 10 shows the most common audio power amplifier.

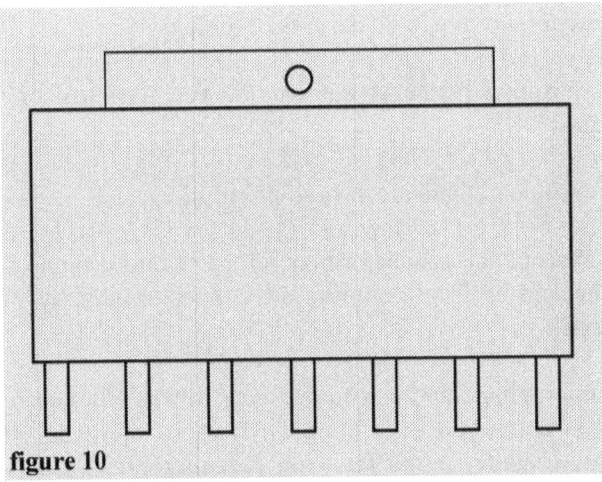

figure 10

In the case where the microphone is suspected of being the problem, a quick and non-technical possible remedy is to spray the sliding push-to-talk switch with a contact cleaner containing a light lubricant. Depress the push-to-talk switch several times while spraying the switch with the cleaner.

Some contact cleaners contain cleansing agents which are damaging to plastics so choosing the proper cleaner is important. Now, if cleaning the push-to-talk switch does not solve the problem, the next step is to check the audio power amplifier integrated circuit (with microphone).

To do this, there are two tests. First, remove the bottom cover of the radio to expose the bottom of the circuit board. Connect the power leads from the radio to the terminals of a power supply of proper voltage and polarity. Now, apply power. Next, turn up the volume control and put your finger on the pins of the audio integrated circuit while being in close proximity of a fluorescent light fixture. If the audio integrated circuit is working, you will hear a very pronounced hum coming from the speaker with the audio volume turned up about halfway. If nothing is heard on the speaker, then the next test is to check the voltage on every pin of the audio amplifier integrated circuit (the audio integrated circuit is usually mounted to the side of the inside frame of the C.B. radio). With the negative lead of a voltmeter attached to the radio's negative power lead, use the positive lead of the voltmeter (using a direct current voltage scale of 12 volts or higher) and measure each pin of the audio integrated circuit. If you have about the same voltage reading on about every pin, it can be almost assured that the audio integrated circuit is bad. If you see a voltage difference by about three to six volts from pin to pin, it can be assumed the audio integrated circuit is good. Further testing should be done by a qualified radio technician if neither of these tests prove to disclose the problem.

8. What is wrong when a signal can be received but one cannot be transmitted?

With this problem, there are several possibilities:

a. The microphone connector may not be securely connected. This can be checked by firmly holding the connector, slightly wiggling it as you push in.

b. The microphone audio is turned way down. The microphone audio level is controlled by a knob on front of the panel of most C.B. radios. Usually, if the knob on the microphone control is turned counter clockwise too far, the microphone audio level is turned down too far. Turn the knob clockwise to increase the level.

c. It is also possible the microphone element is bad. To test it, connect one lead of the microphone element to the vertical input of an oscilloscope and set the volt per division on about one millivolt scale and put the other lead of the microphone to oscilloscope ground. Whistle into the microphone element and notice voltage variation on the oscilloscope. If none is present, the element is bad.

d. The final power transmit amplifier transistor is bad. The test to see if the C.B. radio transmitter is functioning, is to place a number 47 bulb with a PL-259 fitting on the radio's SO-239 antenna's connector, apply power to the radio and key up the microphone. If the bulb illuminates brightly, the transmitter amplifier is good. If the bulb does not, either the final or the driver amplifier may be bad. There is also a possibility that voltage going to either or both of these amplifiers is absent. In order to make the determination of the specific problem, some complex radio troubleshooting diagnostic testing would need to be done which only a qualified radio technician can do.

e. The carrier power is too high. If the carrier power is increased without increasing audio power, the carrier power will tend to "swamp" out the audio power signal. This unbalanced situation will result in low microphone audio output on the carrier. Both the carrier power and the audio power must be increased proportionally for optimum performance.

f. There is no power supplied to the final and/or driver amplifiers for the transmitter. Short of extensive troubleshooting diagnostics, the best one can do is to visually check the circuit for broken or burned circuit connection on the main circuit board of the C.B. radio. An attempt may be made to solder a "jumper wire" over any broken circuit connection. Be certain that a properly rated fuse is in place before power is applied to the C.B. radio.

   In many cases where a circuit connection on a circuit board is found to be burned, a circuit component is likely to be found shorted. At this point, it is advisable to refer the problem to a qualified radio technician.

9. Why can I transmit a signal but cannot receive one?

Again, there are several things that can cause this problem to exist.

a. The R.F. gain control is turned down too far. There is a knob located on the front of most C.B. radios marked R.F. gain. This controls the radio's ability to receive a signal. If the knob is turned in a counter clockwise position, the receiver is "desensed" or has little ability to pick up a weak signal but if the knob is turned clockwise to its maximum, it is in the best position to receive a weak signal.

b. The squelch is turned up too high. On the front panel of most C.B. radios is a squelch knob. The squelch is used to mute any receiver noise until a strong signal is received. If the squelch is set too high, the receiver is muted and too much signal will be needed to enable the receiver to operate.

c. There is a misalignment of the radio receiver. This happens sometimes when a C.B. radio is taken to a C.B. shop for servicing and the technician inadvertently turns the wrong coil or does something to the circuit that will inhibit the receiver to receive well. This is usually a simple problem for the technician to correct.

d. The first radio frequency amplifier in the receiver is bad. When a radio signal is detected by a receiver, it goes through an amplification process to intensify the weak incoming signal. There will be several stages of amplification used in the process of intensifying the weak incoming signal. The first stage of amplification is what is referred to as the first radio frequency amplifier or first R.F. Amp. Being the very first stage of amplification makes it vulnerable to electrical noise variations. Therefore, if these electrical noise variations exceed the design characteristics of the transistor amplifier, the transistor will be destroyed.

   The test for the first radio frequency amplifier, being good or bad, is complex and the recommendation is to have a qualified radio technician perform the tests.

e. There is a bad mixer crystal in the receiver. Older C.B. radios have several crystals that are used to develop, transmit and receive signals. This is done by a process known as Heterodyning, which is mixing two frequencies from two crystals (refer to section IV, page 33). If one or both crystals are off frequency, the developed transmit and receive signals will also be off frequency. When this occurs, the results will be poor or no signal reception. To correct the problem, one or both crystals will need to be replaced. The test used most often is the use of a frequency counter. This again is a complex problem which only a qualified radio technician should perform.

b. The squelch is turned up too high. On the front panel of most C.B. radios is a squelch knob. The squelch is used to mute any receiver noise until a strong signal is received. If the squelch is set too high, the receiver is muted and too much signal will be needed to enable the receiver to operate.

10. What causes a radio receiver to go dead?

There are many things which can cause a radio receiver to go dead and if this occurs, the most prudent thing to do is allow a radio technician to troubleshoot the problem. The odds of a non technical person finding the cause by trial and error is remote. A lot of money can be spent by randomly replacing parts to find the cause only to realize that this trial and error method is costly without finding the solution and in the end, a radio technician must be consulted.

Sometimes a radio receiver can go dead because a driver next to you keys up his radio and causes your radio receiver to go dead. Typically, the driver next to you was running a great deal of power and because your antenna and his antenna were reasonably close together, the high energy emitted from his antenna was "coupled" to yours, sending that energy to your C.B. receiver's first stage of amplification. This transistor amplifier is not stout enough to handle all that energy and that first stage of amplification is usually damaged in the process.

11. Several people I have talked to indicate that my radio sounds bad; however, it seems to sound okay to me on my talk-back speaker.

Talk-back has absolutely nothing to do with the transmitter of your radio. Talk-back will only let you know whether your microphone is working or not working and nothing more. Therefore, it is possible you could have a transmitter problem even though your talk-back is working.

12. As I increase speed going down the highway, the noise level on my C.B. also increases and I seem to get more noise than signal.

Additional static electrical charge buildup occurs on an antenna with the increase in speed. One way it can be reduced is to DC ground the antenna. DC grounding incorporates the use of a coil, going to the ground from the antenna. Any inherent electrical noise will be shunted to the ground by the coil, while at the same time, the outgoing transmitted and receiver signal stays above the coil. Refer to figure 11 below and take note of the caution spelled out in the illustration.

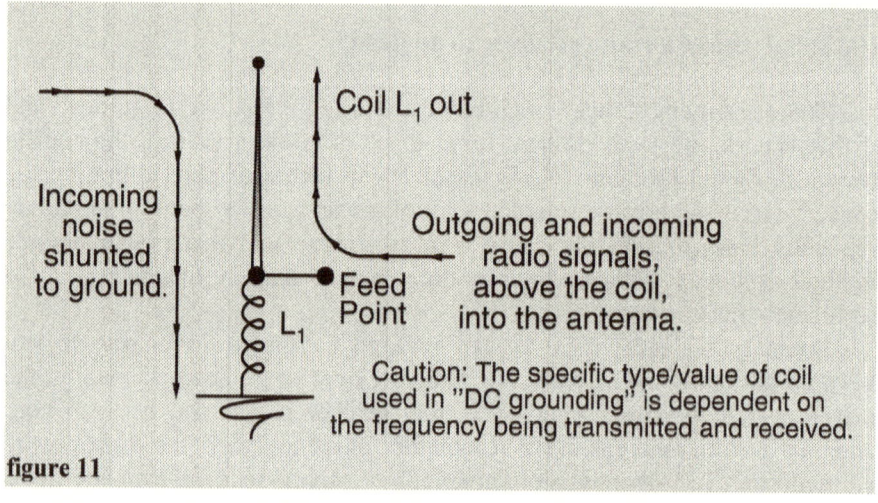

figure 11

13. Every time I key the microphone, the radio squeals.

A couple of things can cause your radio to squeal.

a. The acoustics from the speaker of your receiver having talk-back will feed back into the microphone causing a squeal. One solution is to keep your radio and/or remote speaker (if you have one) separated as far as possible from the microphone. Also check frequently for broken or chaffed wires on the microphone cable.

b. Sometimes, wires in the microphone cable get broken and touched. Then when you key your microphone, you will get a squeal. Periodically, you should check the wires to see if they are broken or chaffed.

14. What is wrong when everytime I hit a bump, my radio quits?

More often than not, when you hit a bump and your radio quits, the C.B. radio's printed circuit board may have a bad electrical solder or friction connection (see figure 12 of printed circuit board on page 27). Sometimes a bad electrical connection is obvious and sometimes not. Troubleshooting intermittent radio problems requires time and a lot of patience.

figure 12

One way to isolate intermittent radio problems is to subject the C.B. radio circuit board to very warm and very cold temperatures. Here is a step by step troubleshooting procedure for intermittent C.B. radio problems:

a. Expose the C.B. radio's printed circuit board by removing its bottom cover (see figure 12 above).
b. Connect power source and antenna (or dummy antenna) to the radio.
c. Apply power to the C.B. radio. Set the volume control to about one-third of its rotation. Set the radio to an active channel.
d. If noise is present and no voice audio from a radio signal is present (or nothing at all is heard), then continue with the following steps.
e. Begin step five by purchasing the following test procedure items:

1. Hand-held hair dryer.
2. One can of component coolant (from most electronic supply stores).

f. Disconnect power source and antenna from the C.B. radio.
g. Remove the bottom cover of the C.B. radio to expose the radio's printed circuit board. Reconnect the power supply and antenna to the radio (microphone connected).
h. Apply power to the radio. Using a hair dryer set at the medium heat level, use a circular motion moving the hair dryer inward and outward toward the center of any point on the printed circuit board. Be sure and keep the hair dryer six or seven inches from the printed circuit board as too much heat will damage the board.

An adequate test of temperature to the radio's printed circuit board is to place your fingers on the printed circuit board being warmed. The circuit board should be just very warm to the touch on your fingers; however, not so warm that you cannot keep your fingers comfortably on the printed circuit board for about five to ten seconds.

Connect the radio to an antenna and power supply. Apply power.

If the radio comes "alive", that is, begins to transmit or receive or noise is heard when heat is applied, then a possible conclusion is a bad solder connection or electronic component is bad.

To further the test after the heat is applied to find out what part is causing the problem, the component coolant will be used. First, however, put on proper insulated gloves and protective eye glasses as component coolant is extremely cold and can cause very serious injury to your body.

Spray the coolant on the suspected area and if the radio goes dead, you have isolated the problem. Trial and error, using the coolant on the printed circuit board, is a way of troubleshooting the circuit to its specific problem. As mentioned earlier, this kind of problem takes time and patience to remedy.

The above mentioned procedure used to isolate intermittent radio problems is specific to a temperature related problem. In the above example, the radio was inoperative at or below room temperature and when heat was applied, the radio began to function properly. The reverse may also be true. That is, the radio may not function at room or higher temperature; however, at lower temperature the radio functions properly.

The procedures for troubleshooting intermittent radio problems when the radio is inoperative at room or higher temperatures are as follows:

Verify the radio has an intermittent temperature dependent problem by placing the radio in a cold environment for about twenty or thirty minutes. A freezer would be better than a refrigerator; however, a very cold refrigerator could suffice.

After removing the radio from the cold environment, connect power, microphone and antenna to the radio. Apply power. If the radio comes "alive", then a reasonable conclusion would be that the radio has a temperature dependent problem.

Now, warm the radio to room or higher temperature with a hair dryer using the same precautions stated above to the point where the radio ceases to operate.

Apply power to the radio. Using component coolant with the same precautions stated above, spray the suspected problem area on the printed circuit board. If the radio begins to function after the coolant is applied, a conclusion of a temperature dependent problem exists. By trial and error, repeat the above steps, only this time, spray a suspected component or certain area of the printed circuit board with coolant until, once again, the radio begins to function properly. It should be evident, at this point, that there is either a defective component, a circuit board break, a bad solder joint or a bad friction connection.

i.   Another test for localizing intermittent radio problems is the circuit board "flex test". This test involves the following steps:

    Remove the bottom cover of the radio to expose the printed circuit board.
    Connect power and antenna to the radio.
    Apply power. With power applied, apply a slight pressure with your fingers on the circuit board to various areas, paying particular attention to duplicate the problem, i.e. the radio receiver or transmitter goes on or off as you are applying pressure and flexing the circuit board. Once you have established intermittent operation of the radio using the flexing of the circuit board test method, localize the specific area of the problem by applying slight pressure on that specific area where the radio ceases or begins to function. Be certain not to cause damage to the board or circuit connection traces by applying too much pressure.

j.   Another simple test for finding intermittent C.B. radio problems is the microphone cable stretch test. The procedure for this test is as follows:

Connect power to the C.B. radio. Increase the receiver volume by rotating the volume selector to about one-third its rotation. If the radio fails to have receiver volume, stretch the microphone cable about one-half the microphone cable length. If the radio receiver volume comes on during the period the microphone cable is being stretched, then there is a break in the cable and the cable should be replaced. Similarly, if there is no transmit function when the microphone is depressed, once again stretch and release the microphone cable. The microphone cable needs to be replaced if the transmitter begins working after the microphone cable is stretched.

A check of the signal on the RF meter on the front of most C.B. radios will indicate whether or not the receiver and transmitter are functioning during the time the microphone cable is being tested.

15. After I had my radio "peaked", I discovered my transmitted audio was lowered and I need to correct this.

When you peak a radio transmitter, you must also provide for additional increase in modulating power (modulation is superimposing an audio signal, i.e. microphone audio, on an R.F. carrier).

Look at it this way: if you are standing on an elevated platform and someone at ground level throws you a ball to catch, it takes a certain amount of energy from that person who is throwing the ball to get it to you. Now, if you elevate yourself on a higher platform, that same person is going to expend additional energy to throw the ball harder in order to reach the higher level where you are.

It is the same idea with peaking a radio. The carrier in this case being the height of the platform and the modulating power is being the energy expended in throwing the ball. (Refer to section VIII, page 70, question 22)

16. Should it be, whenever I transmit, that the transmitted audio signal comes through on my A.M.-F.M. radio?

When this happens, the transmitting antenna and/or the power amplifier is located too close to the A.M.-F.M. receiving antenna and/or radio. The closeness of a radiating device such as an antenna or power amplifier is referred to as "coupling". Reducing the "coupling" effectively reduces transmitted radiated energy into an A.M.-F.M. radio. "Coupling" is reduced by greater separation of the two antennas or the device emitting the radiated energy.

Sometimes it is necessary to place a metal (preferably copper) shield over the A.M.-F.M. radio to overcome the effects of transmitted radiated energy from the transmitting antenna or radiating device. This is referred to as "shielding".

17. What's wrong when there is neither microphone audio nor receiving audio?

When this problem occurs, one possible reason is the audio integrated circuit (I.C.) power amplifier is overheating. The most common cause of the power amplifier I.C. overheating is the heatsink compound (refer to question 29 on heatsink compound, page 78) is dried out and has broken down to the point it ceases to work properly and allow the heat to transfer from the power amplifier I.C. to a heatsink, in this case, the frame of the radio.

To correct this problem, first remove the power from the radio and disconnect the antenna. Remove the screws on the cover of the radio to expose the audio power amplifier I.C. (refer to figure 10, page 21).

Clean the old heatsink compound off the back of the power amplifier I.C., redress the I.C. with fresh heatsink compound, spreading it liberally on the back of the power amplifier audio I.C. Replace the screws affixing the power amplifier I.C. to the heatsink. Hook up antenna and power and apply power to the radio. Leave the radio on for a period of time to duplicate the intermittent problem. If the radio goes on and consistent audio is heard for 15 minutes or more, the intermittent problem is solved. Replace the cover.

Some other things which could cause this problem to occur are:

    a. Bad solder connections and interconnections.
    b. A bad microphone cable.
    c. There is no power to the audio power amplifier.

18. Why is my radio so warm, particularly after I finish transmitting?

When you are transmitting electrical energy, it consumes power. Power is heat. The device in your C.B. radio used to transmit this energy is a transistor. A transistor consumes power in the process of transmitting electrical energy. The transistor will get very warm during the time it is transmitting electrical energy. To keep the transistor from getting too hot and destroying itself, it is usually mounted on the side of the radio metal frame to "sink" or transfer the heat from the transistor. Heatsink compound is placed between the transistor and frame or finned metal heatsink on the radio before the transistor is mounted to insure a good transfer of heat from the transistor to the heatsink.

There is a point; however, when the heatsink itself will get very warm to the touch. Some C.B. radios are equipped with "thermal shutdown devices" on the final transmitter power amplifier. This device will sense elevated temperature changes and will shut down the transmitter, disabling its use until the power amplifier and heatsink have had time to cool. Shutdown devices are used to protect the transmitter final amplifier from damage resulting from excessive heat.

Usually lower powered transmitters are at less risk of overheating by conduction of heat transferred by the final transmitter power amplifier onto the heatsink. On any radio, however, common sense and good judgement should be exercised when transmitting. That is, transmit for short periods of times thus keeping the radio cooler.

Periodically, place your fingers on the side of the radio for a period of about fifteen seconds. If you cannot comfortably keep your fingers on the side of the radio due to the elevated temperature, discontinue using the transmitter until it has sufficiently cooled.

Excessive heat on a transmitter power amplifier (transistor) will shorten the life and degrade the performance of a transistor power amplifier.

1. What is a Phase Locked Loop (P.L.L.) synthesizer?

A Phase Locked Loop (P.L.L.) synthesizer is just a circuit that, instead of using a different crystals for each channel in the radio, uses only a few, and using them in combinations, generates many different frequencies. This saves space and cost of your radio.

A Phase Locked Loop (P.L.L.) is a type of synthesizer that, instead of mixing different frequencies together from several crystals to derive a final frequency, it uses a Voltage Controlled Oscillator (V.C.O.) and very few crystals. Have you wondered how C.B. technicians change the frequencies in radios, for example, the uncommon frequencies? Following is an overview on how a P.L.L. system works, as well as how frequencies can be changed in most radios.

Let's begin with a P.L.L. diagram and a brief description and discussion about it. First of all, the question is what is a Phase Locked Loop? A good place to start would be to discuss frequency generating in a radio. Early synthesized C.B. radios used a process of "heterodyning" or crystal mixing (mixing two frequencies by adding or subtracting X_1 and X_2 to obtain a third). A simple illustration of this is shown in figure 13 below:

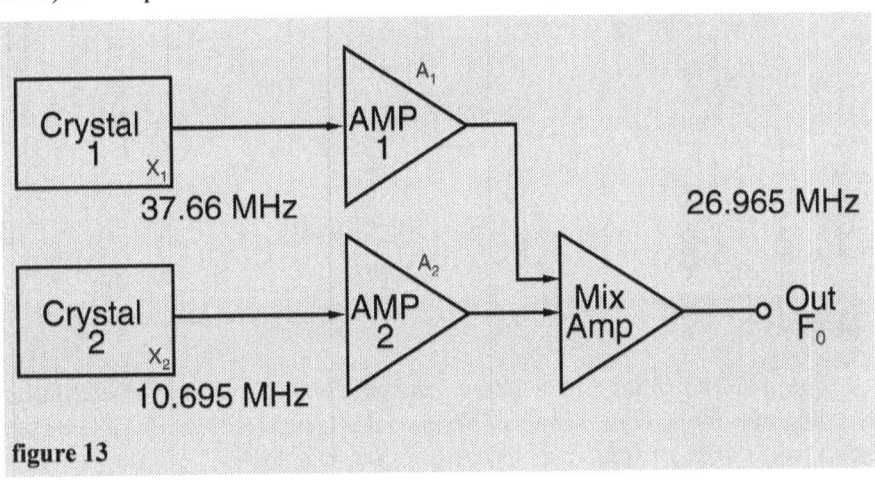

figure 13

In figure 13, we see two crystals, X_1 - 37.66 MHz and crystal X_2 - 10.695 MHz. The signals from both crystals are directed into amplifier A_1 and amplifier A_2 respectively, to increase the signal level of each crystal being generated. These amplifiers then feed a third amplifier called a "mixer" amplifier. At this point, the two crystal frequencies X_1 and X_2 are mixed and the difference or sum of the two is the frequency output F_0. The result is the frequency of 26,965 MHz, which is channel 1 on the C.B. frequency band.

While one crystal can remain the same frequency, another crystal is needed to produce the next selected frequency channel 2.

Clearly, you can see several crystals are needed to complete the process of channel frequency selection up to 40 channels. Crystals are expensive, fragile and tend to drift off frequency in time. The alternative to changing frequencies instead of crystal synthesizing is through the use of Phase Locked Loop. Figure 14, below, is a simple example of a Phase Locked Loop circuit.

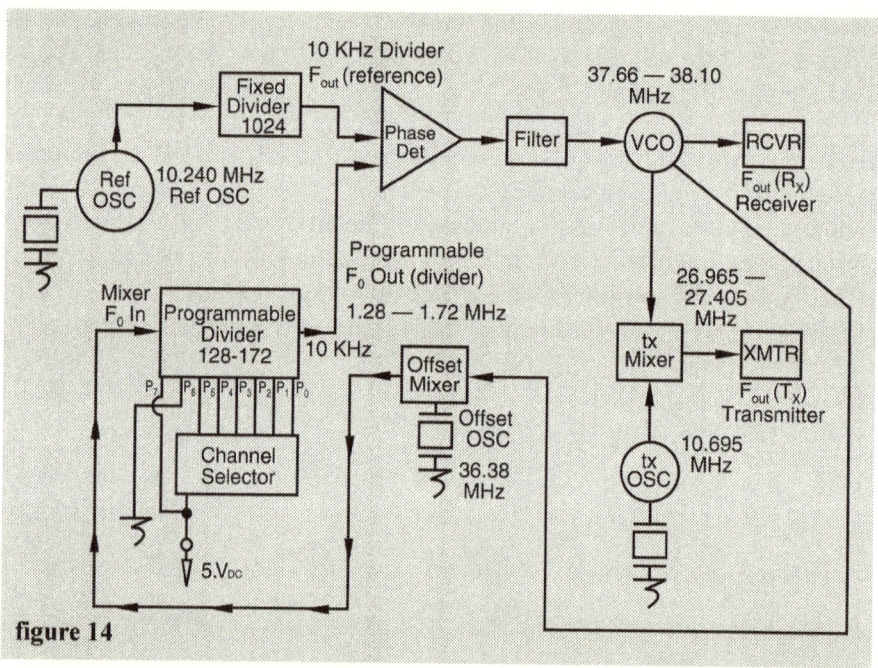

figure 14

The main advantage of a Phase Locked Loop is you don't need but a few crystals for a wide range of frequencies generated. A Phase Locked Loop won't drift off frequency as easily as crystals do.

Phase Locked Loop is less expensive compared to having several crystals for frequency generation.

Now, here is how a Phase Locked Loop works:

The Voltage Controlled Oscillator (V.C.O.) is the major part of the Phase Locked Loop circuit. A V.C.O. is used to generate a signal for various frequencies. A V.C.O. uses a special diode called a varactor. A varactor is a diode that exhibits capacitance and changes when a change in voltage is applied to its terminals. When different voltages are applied to a varactor, capacitance changes and different frequencies will occur. There are charts given for every varactor diode that denote the capacitance of the diode in relation to applied voltage.

The V.C.O. sends its signal to what is known as an offset mixer. These signals combine in the mixer and the result is the difference or sum of these two signals. The offset mixer is used to breakdown the signal from the V.C.O. since the programmable divider is not capable of breaking this V.C.O. frequency down far enough. The signal is now sent to a programmable divider. The programmable divider can be programmed as shown in figure 15 below:

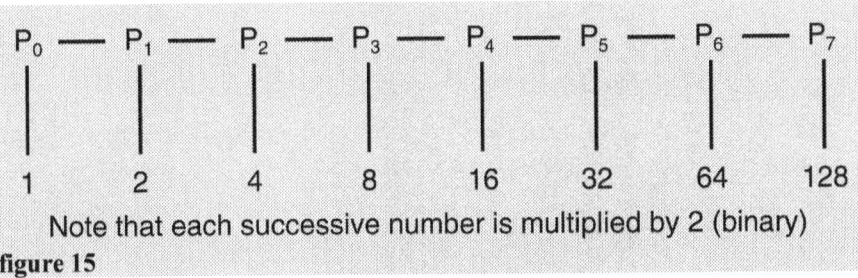

Note that each successive number is multiplied by 2 (binary)

figure 15

In our example, notice the programmable divider and its respective pins P_0 through P_7. Each one of these pins has a specific binary 2 numerical value. The total of numerical value of pins P_0 through P_7 is equal to the sum of all pins values.

$$P_0+P_1+P_2+P_3+P_4+P_5+P_6+P_7 = P_{TOTAL} \text{ or}$$
$$(1+2+4+8+16+32+128) = 191$$

P_6 is at ground and not used.

Now, in order for programming to take place, any combination of these pins may be connected to either + 5 volts (binary Hi = 1) or connected to ground 0 volts (binary Lo = 0).

The programmable divider's total numeric value is equal to the sum of the individual pins connected to a binary Hi (1).

For example:

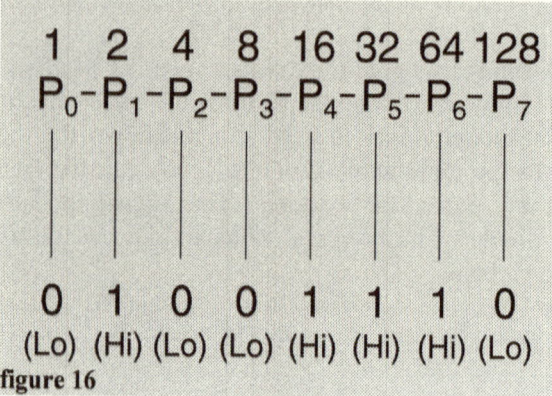

figure 16

The total numeric value of the binary programmable divider recognizing a binary (Hi) on the following pin numbers:

$$P_1 = 2, P_4 = 16, P_5 = 32, P_6 = 64$$ Now we add the numeric value of these pins: $(2+16+32+64=114)$

So now, our programmable divisor will be 114. Now, assume we have a frequency of 1.28 MHz coming from the offset mixer and into the programmable divider. $F_{O(IN)}$ mixer is the incoming offset mixer frequency. $F_{O(IN)}$ mixer = 1.28 x 10^6 MHz to the programmable divider and is now divided by 114, the divisor we just programmed. Now we want to get the frequency coming out of the programmable divider. So frequency out is

$F_{O(OUT)}$ divider = $F_{O(IN)}$ mixer divided by the divisor = 1.28 x 10^6 divided by 114 = 11, 228 Hz or 11.228 KHz.

Now that we have found out what frequency is coming out of the programmable divider

$F_{O(OUT)}$ divider = $F_{O(IN)}$ mixer divided by the divisor = 1.28 x 10^6 divided by 114 = 11, 228 Hz or 11.228 KHz., what happens next and where does this signal go?

$F_{O(OUT)}$ divider (output frequencies from programmable divider) is now sent to a phase detector. This phase detector has two signals coming into it:

a. The signal from the output of the programmable divider.
b. A 10 KHz reference signal coming from a fixed divider.

36

The programmable divider's output ($F_{O(OUT)}$ divider) is now compared to that of the divider's output reference signal frequency of 10 KHz. If the ($F_{O(OUT)}$ divider) programmable divider does not equal $F_{O(OUT)}$ reference, the fixed divider's output, 10 KHz, then the two signals are said to have phase difference. Phase is defined as the amount of electrical degrees one wave leads or lags another. See figure 17 for an example of wave leads and lags.

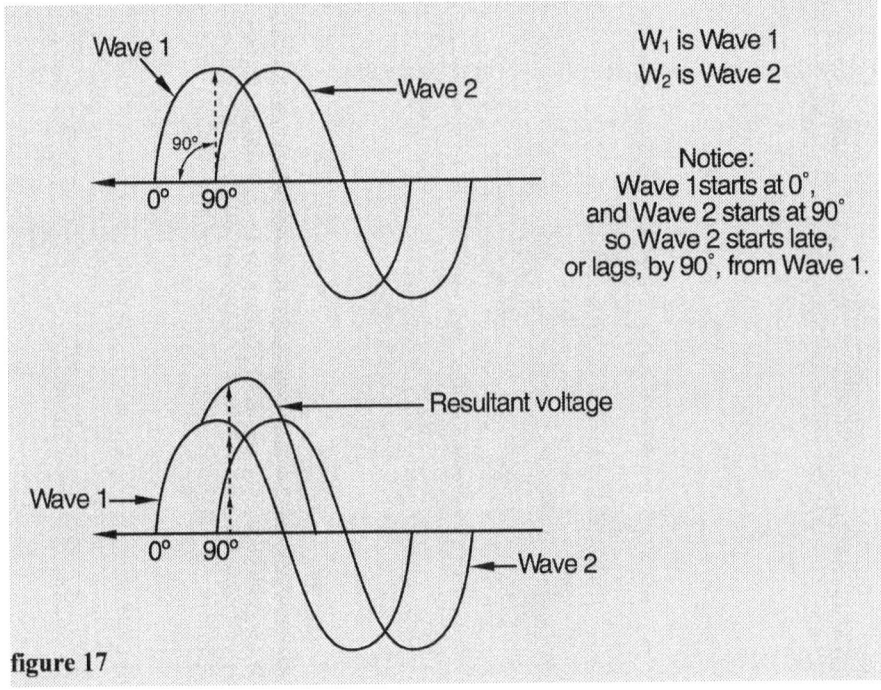

figure 17

Wave 2 (W_2) is said to be lagging Wave 1 (W_1) by 90 degrees. They are both "out of phase". Since they are out of phase, the resultant voltage created by these 2 waves at any point in time are now added together to yield a "resultant voltage".

So now, if there is a phase difference, it will be detected by the "phase detector", and this phase detector's output is an unfiltered D.C. (Direct Current) output going to a filter. The output from the filter is a clean D.C. voltage that is sent back to the Voltage Controlled Oscillator (V.C.O). Changes of voltage on the V.C.O. result in changes in frequency. This frequency is sent back to the offset mixer, programmable divider, phase detector, filter and back to the V.C.O. and the "loop" is complete. This loop is "locked" when no changes occur in phase differences in the phase detector from the programmable divider and fixed divider.

When the loop is stabilized, the frequency coming out of the V.C.O. will feed both receive and transmit circuits. Clearly, we can see the V.C.O.'s output frequency being fed to a transmit (TX) mixer and mixed with a 10.695 transmit oscillator reference frequency and the difference between the V.C.O. frequency and the transmit (TX) oscillator frequency will yield the transmitter's final output frequency.

Phase Locked Loop circuits are usually in the form of an integrated circuit. There are specific charts that give programming and design for every Phase Locked Loop integrated circuit. See figure 18 below:

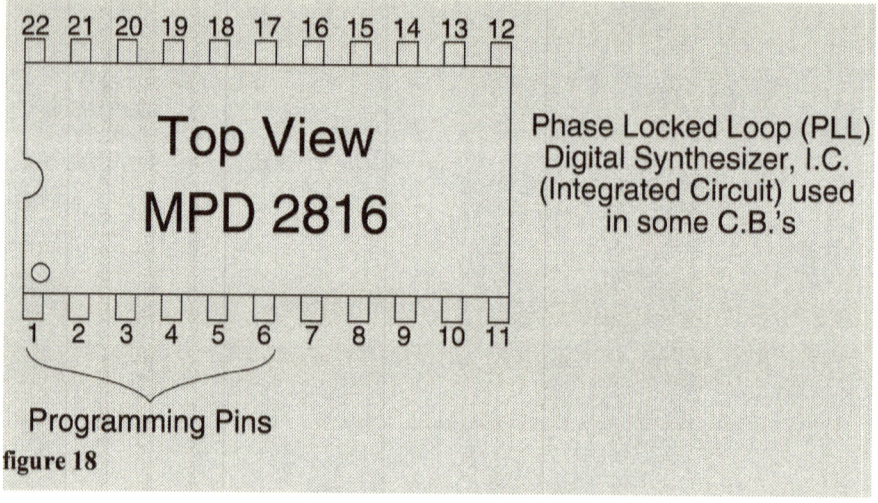

figure 18

Pins 1 through 6 are where you program the synthesizer on this integrated circuit. Each Phase Locked Loop integrated circuit has different programming pin numbers. Note the dot on the integrated circuit right above the first pin (1). This dot is used as an index to begin the numbering sequence of the integrated circuit's pins, beginning below and to the right of the dot.

Also, on some integrated circuits, there will be a number 1 placed just above the integrated circuit's first pin instead of the dot. Either way (number or dot), the numbering sequence of the integrated circuit's pins are the same.

By now, your curiosity should be to the point where you ask, "How can I program my C.B. radio for certain frequencies?" Let's take another example on exactly how certain frequencies can be programmed.

Refer to figure 14 diagram of a Phase Locked Loop circuit. Remember the programming pins P_0 through P_7?

Recall $(P_0 = 1, P_1 = P_2 = 4, P_3 = 8, P_4 = 16, P_5 = 32$
(when not tied to the ground) $P_6 = 64, P_7 = 128$

Now, notice the transmit oscillator frequency is at 10.695 MHz. Notice this 10.695 MHz signal is going to transmit mixer (Tx mixer) where it combines with the V.C.O. frequency for a difference in frequency. Mathematically stated: F_{OUT} (transmitter) = (Frequency Voltage Control Oscillator) - (Frequency Transmitter Oscillator). F_{OUT} (transmitter) = (F_{VCO} - $F_{OSC(TX)}$). So let's suppose you wanted an F_{OUT} (transmitter) = 26.955 MHz.

Let's put this value into our equation of F_{OUT} (transmitter). F_{OUT} (transmitter) = (F_{VCO} - $F_{OSC(TX)}$). 26.955 MHz = (F_{VCO} - 10.695 MHz) by transposing 10.695 MHz, we can solve for the F_{VCO}. This is F_{VCO} = (26.955 MHz + 10.695 MHz). F_{VCO} = 37.65 MHz.

Now that we know the V.C.O. frequency, notice that the signal on Figure 14 is going directly to the offset mixer. The offset mixer has a 36.38 MHz oscillator combined with it. It "mixes" the V.C.O. frequency with offset oscillator frequency that results in a difference in frequency. Stated mathematically:

F_{OUT} (offset mixer) = F_{VCO} - F_{OSC} (offset mixer).

On the preceding page, we calculated our V.C.O. frequency to be 37.65 MHz. Again, we know the offset oscillator frequency to be 36.38 MHz. We now have enough information to calculate the output frequency from the offset mixer F_{OUT} (offset mixer).

F_{OUT} (offset mixer) = F_{VCO} - F_{OSC} (offset mixer)
F_{OUT} (offset mixer) = 37.65 MHz - 36.38 MHz = 1.27 MHz

Now that we have found the output frequency of the offset mixer F_{OUT} (offset mixer), we must now find a numerical value for the channel selector $P_{D(N)}$ = F_{OUT} (offset mixer) ÷ 10 KHz where 10 KHz is the Phase Locked frequency output of the programmable divider F_{OUT} (divider). F_{OUT} (offset mixer) is the output frequency from the offset mixer and $P_{D(N)}$ = Programmable Divider Number. Previously we found the frequency output of the offset mixer to be F_{OUT} (offset mixer) = 1.27 MHz.

We now have enough information to find the programmable divider number $P_{D(N)}$.

$P_{D(N)}$ = F_{OUT} (offset mixer) $\div$ 10 KHz = 1.27 MHz $\div$ 10 KHz = 127.
$P_{D(N)}$ = 127. This will be our divider number.

Remember again, the numerical value of the programmable divider on the P.L.L. circuit.

$P_0 = 1$, $P_1 = 2$, $P_2 = 4$, $P_3 = 8$, $P_4 = 16$, $P_5 = 32$, $P_6 = 64$, $P_7 = 128$

So what combination (P_0 - - - - P_7) will add up to the numerical value of 127, to equal our $P_{D(N)}$? If you said
($P_{D(N)}$ = $P_1 + P_2 + P_3 + P_4 + P_5 + P_6$)
you are correct. Those pin numbers will then be tied to a +5 volt source.* Now, what number will be tied to ground (out of the circuit)? It is $P_7 = 128$. You should have a good picture on how this works.

Briefly, pins P_0 - - - - P_6 are said to be HIGH (tied to a +5 volt source). Pin P_7 is said to be LOW (tied to the ground). Let's briefly summarize the steps needed to get certain frequencies programmed into a C.B.

1. Find F_{OUT} (transmitter) = F_{VCO} - F_{OSC} (transmitter)

2. Find F_{VCO} = F_{OUT} (transmitter) + F_{OSC} (transmitter)

3. Find F_{OUT} (offset mixer) = F_{VCO} - F_{OSC} (offset mixer)

4. Find $P_{D(N)}$ = F_{OUT} (offset mixer) $\div$ 10 KHz

Tie appropriate pins from the programmable divider to either 5 high volts or low ground to correspond to the $P_{D(N)}$ (Programmable Divider Number).

If these preceding steps are followed, the radio should operate on the frequency (transmit $F_{OUT(TX)}$) and receive ($F_{OUT(RX)}$) selected.

*+5 volt source, is Direct Current

1. What is sideband? What is the difference between sideband and A.M.?

We defined a citizen band carrier frequency to be 27 MegaHertz (MHz) or 27 million cycles per second. See figure 19 below:

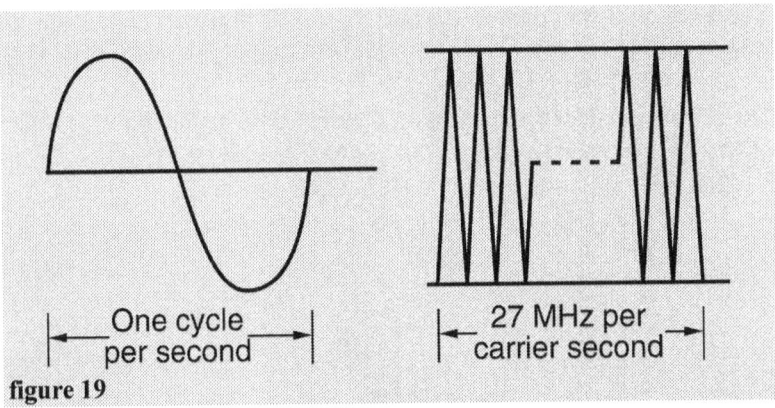

|←— One cycle per second —→| |← 27 MHz per carrier second →|

figure 19

Recall that a cycle of radio signal represents the number of times a voltage waveform shapes and repeats itself over a referenced period of time, usually one second.

Suppose we add an audible signal to the carrier frequency. Let's say the subaudible signal is 2½ cycles per second. The resultant signal will look like figure 20 below:

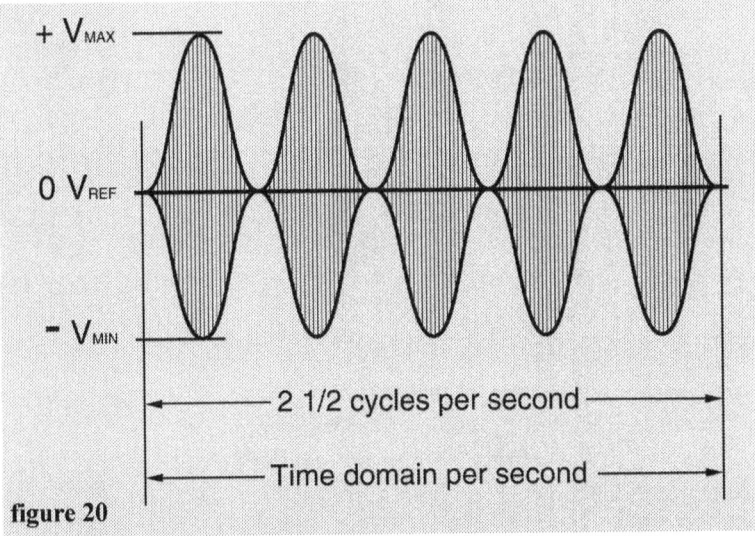

figure 20

You can clearly see that the carrier has been changed to peaks and valleys. This change, due to the addition of an audio signal to the carrier, is called Amplitude Modulation or A.M. Amplitude refers to the intensity of the carrier in volts (+ V maximum - V minimum).

Modulation refers to superimposing an audio signal on a carrier.

Whenever you modulate a carrier, you will always change the carrier frequency waveform. Whenever the carrier frequency waveform is changed, various other frequencies will appear.

These various other frequencies are referred to as sidebands. Sidebands are defined as a frequency above or below the carrier frequency.

For example, suppose we modulate a C.B. carrier 27 MegaHertz (MHz) with a 1, 000 Hertz (Hz) tone. What are the sidebands going to be? We see in figure 21, below, that three frequencies exist.

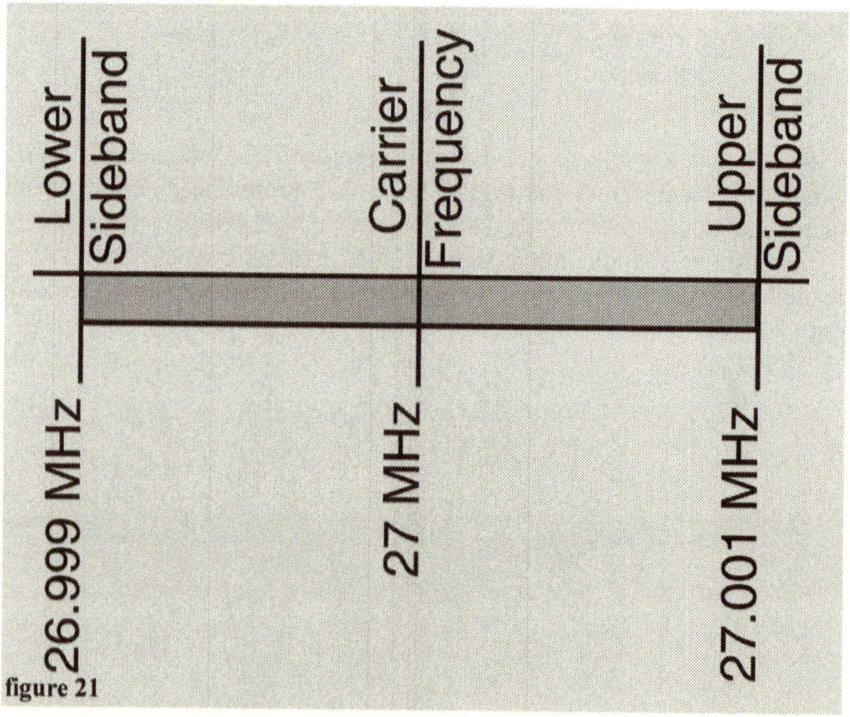

Lower Sideband — 26.999 MHz

Carrier Frequency — 27 MHz

Upper Sideband — 27.001 MHz

figure 21

a. The original frequency (carrier).
b. The other original frequency (audio not shown).
c. The sum of the original frequency and the modulating frequency (upper sideband).
d. The difference of the original frequency and modulating frequency (lower band).

42

Since the other original frequency (modulating audio) is low in frequency, this frequency is usually lost and not transmitted due to a low resistance in the tuning circuit and the antenna. All others, however, are tuned and radiated from the antenna.

A broadband receiver, tuned to accept the range of frequencies of upper and lower sidebands, can be tuned to either upper or lower sidebands and recover the modulated signal, in this case, 1,000 Hz tone.

Suppose we wanted to receive the lower sideband. Using a receiver, the lower sideband and the carrier frequency would be tuned. This is known as lower sideband, tuned carrier or LSB-TC. Similarly, the same procedure is followed if we want to select the upper sideband.

Since sideband power is higher with transmitted carrier, the consideration of power dissipation is introduced. One way to reduce power dissipation is by suppressing the carrier.

Until there is modulating audio present, there is no carrier because it is suppressed or eliminated. The total power produced by the carrier in the upper and lower sidebands (when modulated) is the sum of the power in the upper and lower sidebands plus the carrier power. Here are the parts of the equation used in calculating sideband power:

Total Power = P_{TOTAL} = $P_{UPPER\ SIDEBAND}$ + $P_{LOWER\ SIDEBAND}$ + $P_{CARRIER}$
Carrier Power = $P_{CARRIER}$
Upper Sideband Power = P_{USB}
Lower Sideband Power = P_{LSB}
P_{TOTAL} = $(P_{USB} + P_{LSB} + P_{CARRIER})$

Example "A" - If we wanted to find out the power of, say, the upper sideband (P_{USB}), we could, by transposing the equation:

P_{TOTAL} = $(P_{USB} + P_{LSB} + P_{CARRIER})$
P_{USB} = $(P_{TOTAL} - P_{LSB} - P_{CARRIER})$

Example "B" - Suppose we wanted to transmit the upper sideband (USB) plus the carrier. By substitution in the P_{TOTAL} equation, we have.

$P_{TOTAL} = (P_{USB} + P_{LSB} + P_{CARRIER})$ solving for $P_{USB} + P_{CARRIER}$, we have:
$(P_{USB} + P_{CARRIER}) = (P_{TOTAL} - P_{LSB})$ therefore, $P_{USB} + P_{CARRIER} = P_{TOTAL} - P_{LSB}$

$$P_{TOTAL} = P_{CARRIER} + \frac{M^2 P_{CARRIER}}{4} + \frac{M^2 P_{CARRIER}}{4}$$

Power Uppersideband
Power Lowersideband

This simplifies to: $P_{TOTAL} = P_{CARRIER} + \dfrac{M^2 P_{CARRIER}}{2}$

If a carrier is modulated 100%, 50% of the total power in that carrier goes into sidebands (25% in the upper half, 25% in the lower half). As the modulated audio decreases below 100%, the transmitted power also decreases. Sideband power can be calculated using the same formula for modulating power (question 22, page 70). Total power for upper sideband and lower sideband is:

$P_{SB} = M^2 P_{DC} \div 2$, where P_{SB} = power in upper and lower sidebands

M = modulation % expressed as a decimal (i.e. .2 = 20%)

P_{DC} = direct current carrier input power to the final power amplifier

Example "C" - Suppose we want to receive the lower sideband (LSB) and suppress the carrier. How would this signal be received if the carrier is suppressed?

The solution: If the carrier is suppressed, the receiver has no way of locking onto a signal that is not present, the signal being the carrier. Therefore, a signal from an oscillator (located in the receiver) is operating at the same frequency as the carrier. The signal is now injected into the receiver. It is mixed with the incoming lower sideband signal, to produce a different frequency that results in an audio frequency signal.

Example "D" - Below illustrates how a lower sideband signal is transmitted and received.

a. Carrier frequency (F_C) = 1 MHz = 1,000,000 cycles
b. Modulating frequence (F_M) = 2 KHz = 2,000 cycles
c. Frequency lower sideband = F_{LSB}

The lower sideband would be the difference of the carrier frequency and modulating frequency. Mathematically $F_{LSB} = (F_C - F_M)$. so:

$F_{LSB} = (1,000,000$ cycles $- 2,000$ cycles$) = 998,000$ cycle or 998 Kilocycles. See figure 22 below:

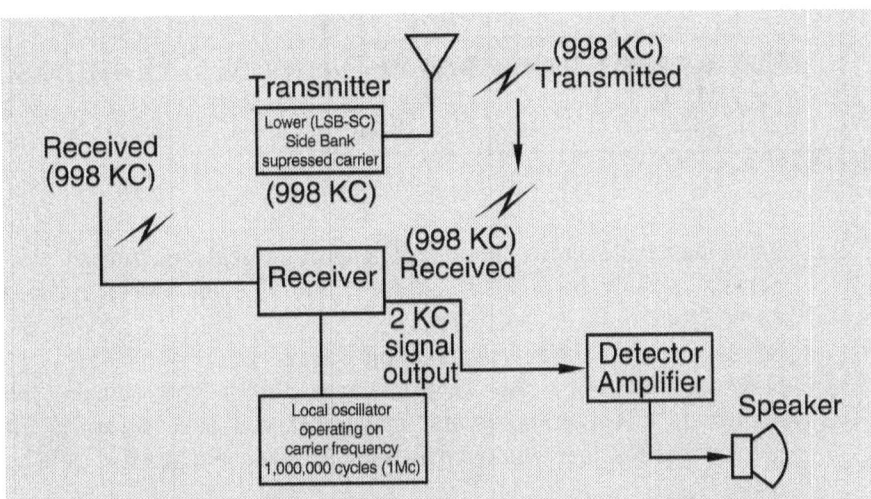

Local Oscillator produces simulated (false) carrier into receiver. Signal is mixed with the incoming received signal and local oscillator, producing a difference output frequency (in this case 2kc) to be detected, amplified, and routed to speaker.

figure 22

Back to example "D". Knowing the lower sideband frequency to be 998 KC and having the carrier frequency suppressed, the above illustrates how the lower sideband is received and carrier is suppressed.

2. How do I use the S/RF meter built into my radio and shown in figure 23, page 46?

When using the meter as an SWR bridge:

a. Key up the transmitter (no voice or tone).
b. Put the switch (located on the front of the radio labeled S/RF, SWR, CAL) in the calibrate (CAL) position.

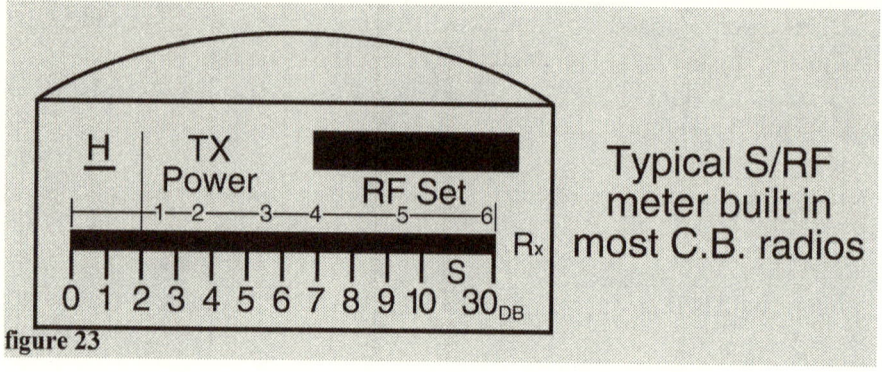

figure 23

c. Place the needle on the meter at the end of the scale marked set. Setting is done by a knob on the front of some radios marked "S.W.R. set".

d. Set the switch in the S.W.R. position.

e. Now read the needle "H". The scale is read from 1-10, 10 being the highest S.W.R. reading (worst match). If you have a reading of greater that 1½, then you have an antenna mismatch and it should be tuned.

f. In the signal (S) or Radio Frequency (RF) position of the (S or RF or S.W.R.) switch, you can now read a received signal (S) level, usually located on the bottom scale. The highest number (usually marked 30 db) provides the best signal while the lower numbers would provide a less quality signal. Take note, however, that many radios do not have a switch for the S position and/or the RF position. When the microphone is keyed, you will read the "RF" or power output on the scale. With the microphone unkeyed, the meter will now read the "S" scale or received signal level scale.

g. Now, place the switch in the RF position and key up the transmitter (no voice). Read the top scale (H) marked RF for power output. This scale will have a range from 0 to 5. An average signal reading will be between 4 and 5. Built-in SWR, S and RF meters are reasonably accurate.

1. Would you identify some of the connectors used in my radio?

Some of the common connectors are:

a. A PL-259 male UHF connector (shown in figure 24) is commonly used on the cable of the transmitting and receiving antenna. It is threaded and plugs into a female connector, usually on the back of a transmitter and/or receiver. UHF refers to Ultra High Frequencies. This ranges from 300 Mega Hertz to 800 Mega Hertz.

b. A SO-239 UHF connector (shown in figure 25) is a female chassis mount connector. It's the counterpart of the PL-259 connector. It is usually mounted on the back of a radio communications transmitter and receiver chassis.

c. A UHF Barrel connector (shown in figure 26) is used to couple together or extend a radio transmission cable by threading and plugging into two PL-259 UHF connectors.

d. The male Motorola connector (shown in figure 27) is used on a radio communication's transmission cable to connect to a radio signal from a receiver to an antenna. This type of connector is used for receiving. You will see this type of connector on the back of A.M.-F.M. automotive radios. This type of connector is usually used on radio receivers.

e. A male RCA (Radio Corporation of America) connector (shown in figure 28) is commonly used for coupling audio signals from an audio amplifier to a speaker. This type of connector will typically plug into a female RCA connector on the back of an audio amplifier chassis and speaker cabinet. The RCA connector has also been used to couple transmitted and received radio signals from a transmitter and receiver to an antenna.

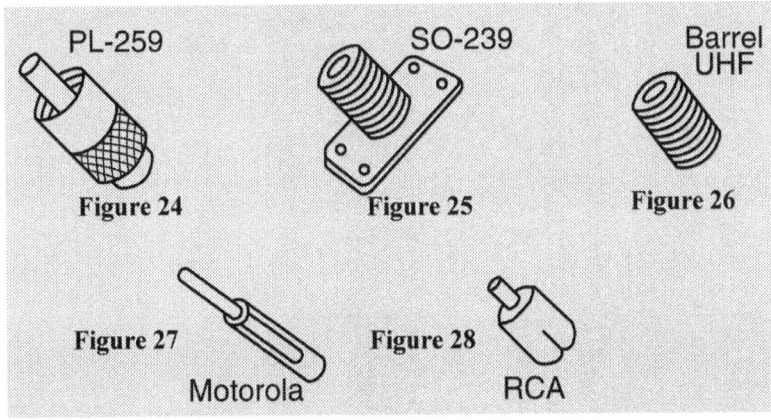

PL-259   SO-239   Barrel UHF

Figure 24   Figure 25   Figure 26

Figure 27   Figure 28

Motorola   RCA

Illustrated below in figure 29, is a typical "L" type C.B. antenna mounting bracket, along with a breakdown of its assembly.

The most common problem encountered while installing this type of bracket is leaving out the insulating fiber, or neoprene washers, and connecting the coaxial cable without them. With the coaxial cable installed without these washers, the radio transmitted and received signals will be shorted out.

It is critical to have all the parts installed, as shown in figure 29 below.

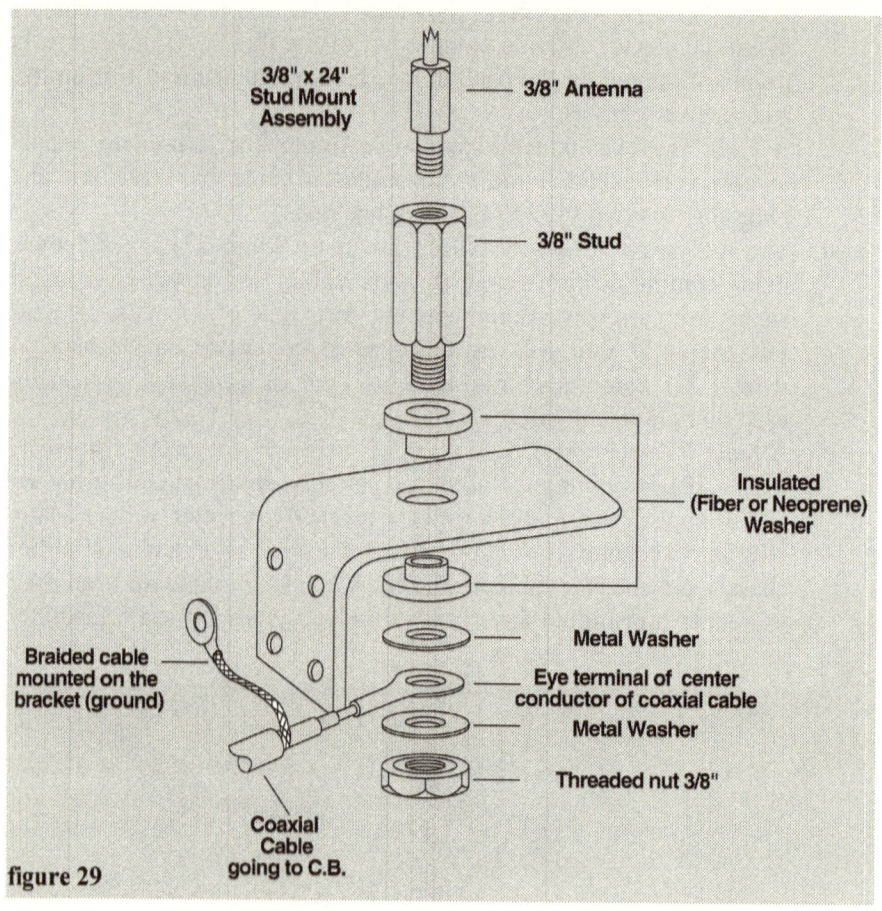

figure 29

1. What is power lead filtering and how is it used?

In terms of radios, power lead filtering is the process of removing an electrical noise component from the power jack to the radio. This could be noise from electric fans, wiper motors, et cetera. Additionally, other types of filters are used to reduce the effect of adjacent frequency interference.

A simple common noise filter is made up of a coil of usually varnished (for insulation) copper wire and a capacitor. A coil of wire is referred to as an inductor, denoted by "L". See figure 30 below:

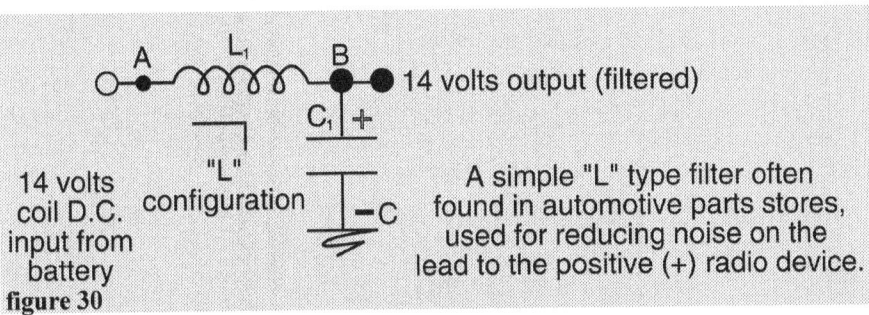

A — L_1 — B 14 volts output (filtered)
C_1 +

14 volts coil D.C. input from battery

"L" configuration

—C

A simple "L" type filter often found in automotive parts stores, used for reducing noise on the lead to the positive (+) radio device.

figure 30

When an alternating current source is applied to the inductor, a changing magnetic field is produced around the diameter of the inductor's wire. This magnetic field moves outward from the wire as current in the inductor wire increases, and inward as current decreases.

When the alternating current source changes polarity, the magnetic field in the inductor collapses into the inductor wire, causing the current to reverse in the wire of the inductor. As the current changes, the collapsing magnetic field will also cause a change of polarity in the inductor. These changes cause the inductor's polarity to oppose the current source polarity.

This opposition to the flow of alternating current is what is referred to as reactance, impedance or alternating current resistance. As the alternating current source frequency increases, so does the reactance of the inductor, and a.c. (alternating current) voltage drop across the inductor "L_1" between points "A" and "B" (figure 30), leaving very little a.c. voltage across the capacitor "C_1" at points "B" and "C". This, ideally, is what we want i.e. blocked a.c. with only a direct current (d.c.) between points "B" and "C". (What little a.c. variation appear between points "B" and "C" will be filtered to ground through Capacitor "C_1".

An inductor has a certain value that is associated with it. It is rated in terms of "Henrys". A Henry is a unit of inductance. So, one Henry equals a change of current of one ampere per second to produce or induce one volt in an inductor.

Another component used in a filter is called a capacitor. A capacitor has the ability to store an electrical charge. Capacitors have various electrical charge storage capacity. This storage capacity is rated in Farads. A Farad is the quantity of electrical charge contained in a capacitor. A Farad is equal to one coulomb, having a potential difference on the plates of a capacitor of one volt. A coulomb is a quantity of electrons, specifically, 1.25×10^{18} electrons (an electron is a negatively charged particle). Similar to an inductor, a capacitor has reactance (alternating current resistance). The difference in reactance of a capacitor to that of an inductor is that a capacitor exhibits a lower reactance as frequency increases, whereas reactance increases with frequency in an inductor.

One reason the reactance is lower in a capacitor is that when frequency increases, the transit time of electron flow moving from plate to plate is faster in a capacitor, thus increasing current flow from the capacitor. As current increases, resistance decreases.

Let's look at an example of an "L" type filter, using hypothetical values:

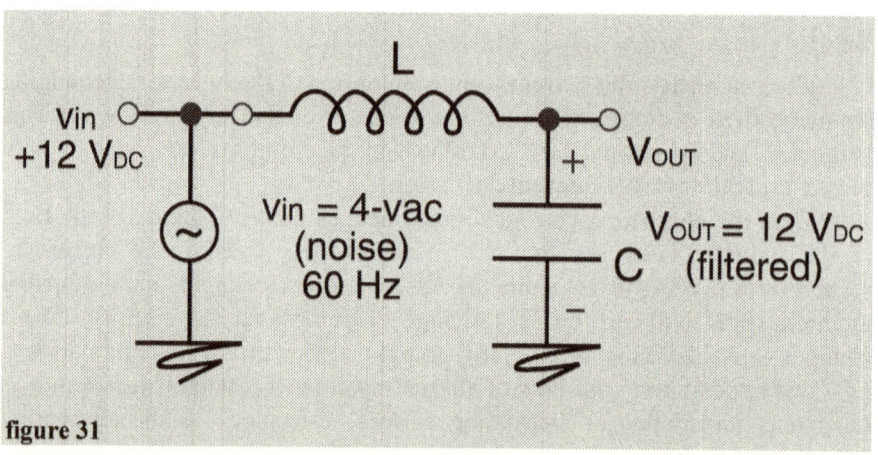

figure 31

Let's assume we have an input voltage (Vin) of 4-Volts a.c. (Alternating Current).

This 4-Volts a.c. is riding on a Direct Current (D.C.) line to the radio devise. The a.c. Component will appear as noise in the device we are operating. Therefore, this 4-Volts a.c. noise we do not want and we strive to "filter" it from our D.C. Supply.

So, let's begin the procedure to design and construct the filter. As in any problem, a solution begins with what is known about the problem. What do we *know* about it?

1. We have 4-Volt a.c. Noise
2. The 4-Volts a.c. is what we want to filter.
3. We want to have as much a.c. Voltage dropped across the inductor "L" as possible and ideally "nothing" appearing (a.c. Wise) at the output V$_{OUT}$. The only voltage we want to appear there is a Pure Direct Current (D.C.). In this case, 12-Volts (what we put into the filter). You can expect a slight voltage drop (D.C.) on V$_{OUT}$, simply because the inductor "L" has some D.C. resistance. Therefore, there will be a slight D.C. voltage drop across the inductor "L".
4. We need mathematical formulas to solve the problem.

What *don't we know* about the problem:

1. The inductance value of "L".
2. The capacitance value of "C".
3. The voltage (a.c.) drop across the inductor "L".
4. The voltage (a.c.) drop across the capacitor "C".

Let's begin by setting up the formulas. A close approximation is...

$$X_C = \frac{.159}{fc}$$ (X$_C$ is a.c. resistance of the capacitor "C" in Ohms)

C = Capacitance of "C" in micro farads (1×10^{-6})
X$_L$ = 2πFL (X$_L$ is a.c. resistance of the inductor "L" in Ohms).
F = Frequency of the noise in Hertz (Cycles/Second).
L = Inductance in Henrys
D.C. Coil (inductor "L") resistance, 3 Ohms.
Now, let's assign a hypothetical value to the inductor to begin.
π = 3.1415 (constant)
L = 40 Henrys, F = 60 Hz (Hertz)
X$_L$ = 2πFL, so X$_L$ = (6.28) (60) (40) = 15072 Ohms
Now, let's assign a hypothetical value to the Capacitor "C".
C = 450 Micro farads (abbreviated Mf) or 450 x 10^{-6} farads

$$Xc = \frac{.159}{FC} = \frac{.159}{(60 \, H_2)(450 \times 10^{-6} \, F)}$$

So, X$_C$ = 5.9 Ohms or 6.0 Ohms.

At this point, we have the following information.
L = 40 Henrys, C = 450 Mfd, X_C = 6.0 Ohms, X_L = 15072 Ohms

This is now enough information to evaluate design of our filter problem. So now, let's evaluate the problem further, finding a.c. (alternating current) voltage drops across the inductor "L" and Capacitor "C".

This evaluation will enable us to examine the filter output voltage, telling us if our filter design has successfully done its job or not of filtering out the unwanted a.c. voltage variation. (In our problem, recall the 4-Volt a.c. noise on the 12-Volt D.C. supply voltage to the filter).

Now, using trigonometric vector analysis, the problem can be solved for voltage drops across inductor "L" and Capacitor "C".

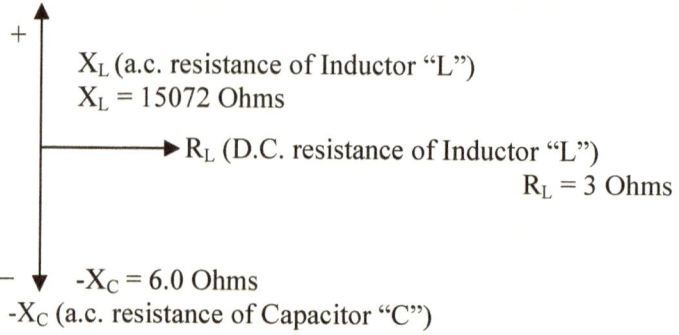

+

X_L (a.c. resistance of Inductor "L")
X_L = 15072 Ohms

R_L (D.C. resistance of Inductor "L")
R_L = 3 Ohms

− $-X_C$ = 6.0 Ohms
$-X_C$ (a.c. resistance of Capacitor "C")

Since there is a 180^0 phase difference between the component of a.c. resistance of the inductor "L" and Capacitor "C", they can be combined algebraically. So, if X_L = 15072 Ohm and $-X_C$ = 6.0 Ohms
$(X_L - X_C) = (15072 \text{ Ohms} - 6.0 \text{ Ohms}) = 15066 \text{ Ohms}$
15066 equals the sign (+) of the highest value, in this case, it is the inductor "L" component of its a.c. resistance.

Rewriting the vector diagram, it follows

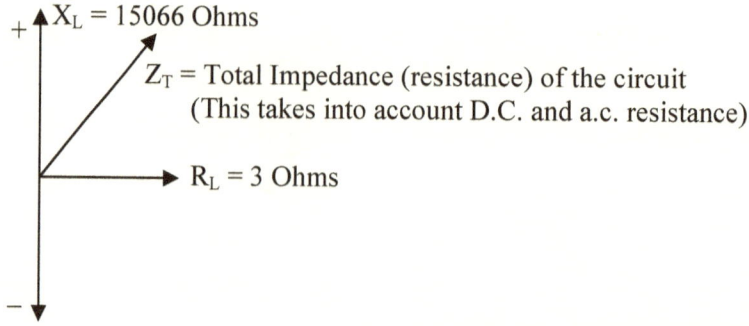

+ X_L = 15066 Ohms
+

Z_T = Total Impedance (resistance) of the circuit
(This takes into account D.C. and a.c. resistance)

R_L = 3 Ohms

−

The result is a right triangle

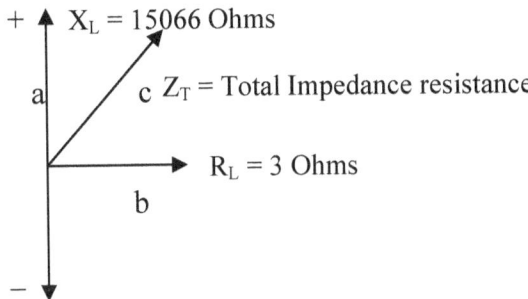

Using Pythagorean Theorem, $a^2 + b^2 = c^2$, total circuit resistance (impedance "Z_T",) (c) can be found. Taking the square root of both sides

$$\sqrt{c^2} = \sqrt{a^2 + b^2} = c = \sqrt{a^2 + b^2}$$
$$c = \sqrt{(15066)^2 + (3)^2} = \sqrt{(2.26 \times 10^8) + (9)}$$

$c = 15066$. Since X_L is much greater than R_L ($X_L \gg R_L$), then X_L predominates. The input/output of the filter circuit appears to be reactive (a.c. resistance). Therefore, $Z_T = 15066$ Ohms.

Using Ohms' law, ($I = V/R$), we substitute the values of V, R to find the total circuit current. I = Current, V = Voltage, R = Resistance. I_T = Total Circuit Current, v = vac = 4 vac (volts a.c.), R = Z_T, total circuit impedance (combined a.c. and D.C. resistance) therefore,

$$I_T = \frac{V_{AC}}{Z_T} = \frac{4}{15066} = .00026 \text{Amps}$$

Having found the total circuit current, we can now (using Ohms' Law) find the voltage drops across the inductor "L" and Capacitor "C".

Ohms' law I = V/R, rewriting the equation, V = IR,

we can now solve for voltage drops across inductor "L" and capacitor "C". Substituting values, into the Ohms' law equation, follows"

I = I_T (Total Circuit Current)
V_L = Voltage drop across inductor ("L")
V_C = Voltage drop across Capacitor ("C")

Substitute X_L = R and X_C = R into Ohms' Law Formula.

First, we choose to solve for "V_L"

$$V_L = I_T X_L = (0.00026) (15066) = 3.917 \text{ Volts}$$
$$V_L = 3.917 \text{ Volts}$$

Now, we solve for V_C

Again, $V_C = I_T X_C = (.00026) (6.00) = .0016 \text{ Volts}$
$V_C = .0016 \text{ Volts}$

At this point, inspection of the Voltage drops across the inductor (V_L) and capacitor (V_C) suggests that almost 4-Volts a.c. is precluded from entering the filters output across capacitor "C" (V_{OUT}).

Now is the time to ask ourselves if we met the objective of filtering out the unwanted 4.0 Volts a.c. The answer is yes. Only 1.6 thousandths of a volt a.c. appears at the output of the filter V_{OUT}. This more than satisfies our filter requirement.

The absence of a load resistance (R_L) of the device being powered, was purposely left out for clarity. Normally, the load resistance (R_L) is greater than the reactance of the capacitator (C), therefore, the effects of R_L can be ignored (see figure 33 on page 55).

To summarize, there are many different variations and types of filters to discuss briefly.

However, the one just presented the "L"-type filter is one of the most common used in noise suppression on power supply circuits.

Filters are designed to stop certain electric noise that occurs at specific frequencies (see figure 32 on page 55). Notice the notation A_V. This is the voltage gain of the noise being produced. Now notice F_C, on the bottom of the illustration. F_C is the "cutoff" frequency of the electrical noise that you wish to stop. Reduction of signal (noise) usually starts at .707 of peak noise voltage and gets progressively lower at specific frequencies.

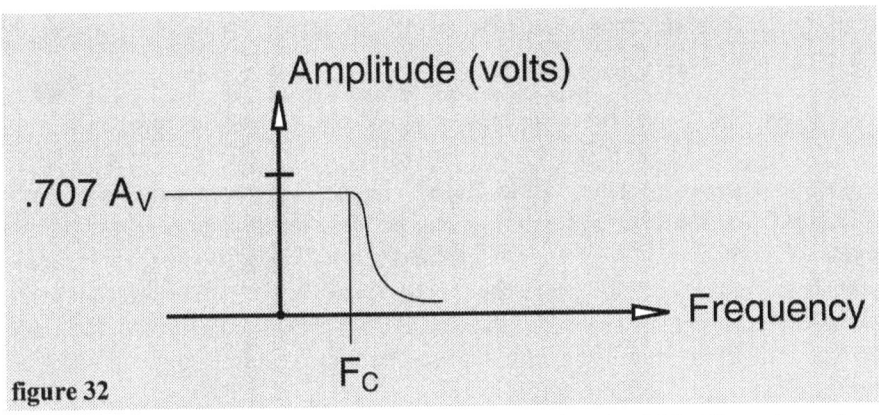

figure 32

In our filter problem example, in order to solve for "C" (Capacitance) or "L" (Inductance), you must determine the frequency of the alternating current (a.c.) noise being generated on the radio power lead.

This can be determined by an oscilloscope discussed briefly. The frequency of the noise you want to eliminate.

To find the frequency of the noise being produced in the radio receiver, you need to determine its cycle per second (frequency). Using an oscilloscope, connect the oscilloscope ground lead to the negative power lead terminal going to the device ground and the oscilloscope vertical input lead (probe) connected to the power lead positive terminal.

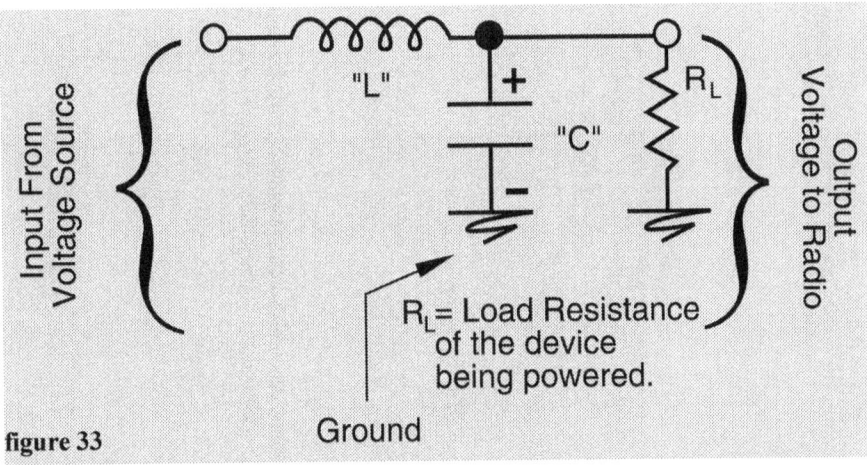

figure 33

Next, display 1 cycle waveform on the gradicule of the oscilloscope. The oscilloscope is a television resembling device with a phosphor screen, marked off in squares. Each square represents 1/100 of a meter or one centimeter. The screen's face, marked off in squares, is called a gradicule (see figure 34 below).

When one cycle of a waveform is displayed on the gradicule, the waveform can be expanded or condensed (narrowed) by means of an adjustment knob marked "Time Base". Time Base represents "Time per Division". Each division is equal to one centimeter. So, for example, if one cycle waveform occupies four centimeters, as displayed on the gradicule shown in figure 34 below, and the "Time Base" is set at one micro second per centimeter per division, the time (period) of this waveform is determined by the following equation: amount of squares one cycle occupies multiplied by time per division. Therefore:

4 x 1 microsecond per division equals 4 microseconds, abbreviated μ sec.

Now that we know the time, 4 μ seconds, we can now find the frequency of this waveform by the following equation:

$F = 1 \div T$

μ = 1 micro or $1 \div 1,000,000$ and M=1 milli or $1 \div 1,000$.

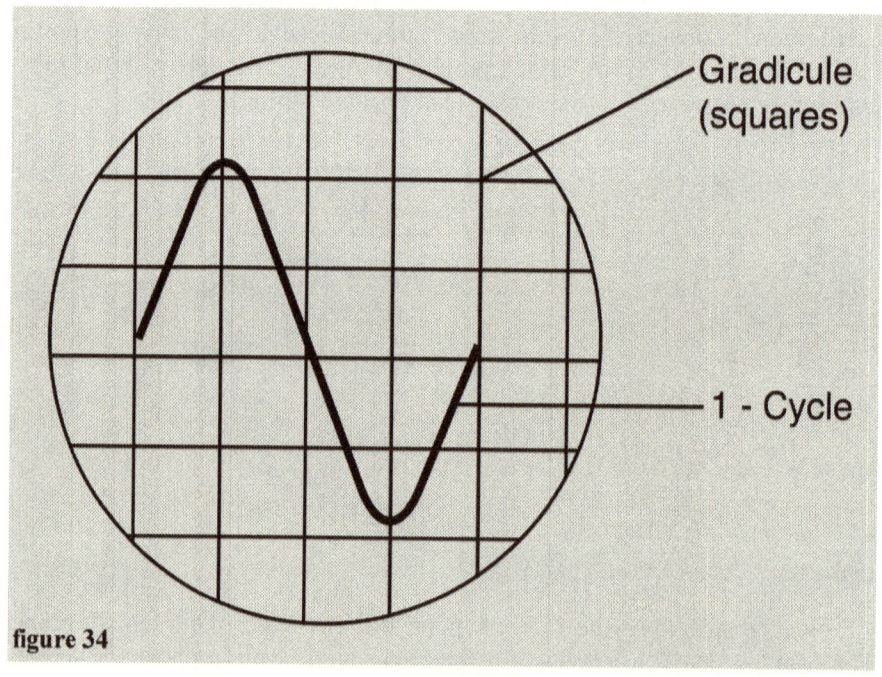

figure 34

Where F = frequency of the waveform and T = time it takes to make 1 cycle waveform.

Then it follows, putting in the values, that
F = 1 ÷ 0.000004 = 2,550,000 cycles per second.

1. What is a carrier?

In radio terminology, a carrier is an electric mobile means for voice, data or other information to be carried through free space.

Anytime an event occurs, for example, a light flashing on and off over a period of time is referred to as frequency. Similarly, if a sinewave were to shape and repeat itself over a period of time, usually one second, this also is a frequency of occurrence.

Now we may ask, what is radio frequency (R.F.)?

Radio frequency is defined as usually having thirty thousand (30,000) or more positive and negative voltage events (or cycles) happening in a period of one second of time. A carrier is comprised of cycles of varying voltage (positive to negative and negative to positive per one second of time).

A carrier will usually contain thirty thousand (30,000) or more cycles of continuously varying voltage (referred to as an R.F. carrier). When another frequency, usually much lower than a carrier frequency, is superimposed on the carrier, a combining effect of the two signals will occur which is called modulation. The carrier is directed to a radiating element called an antenna. An electromagnetic and electrostatic signal is then radiated in free space and when modulated with another signal, the carrier becomes an electric mobile means for voice, data and other information to be carried through free space. Hence, the name "carrier".

2. What is modulation and demodulation?

Modulation is the process of superimposing an audio signal anywhere from about fifty (50) to eighteen thousand (18,000) cycles on a carrier. Demodulation is the process of removing the audio signal from the carrier signal.

a. What is an audio signal?

An audio signal is an audible sound (in the form of voltage varying at an audio rate). Audio signals vary from about fifty (50) cycles per second of time to about eighteen thousand (18,000) cycles per second of time.

b.   How does audio relate to R.F. carrier?

The carrier will vary in voltage (voltage intensity is called "amplitude") at whatever modulating voltage (audio) cycles are applied to the carrier.

3.   What is receiver sensitivity?

Receiver sensitivity is the ability of a radio receiver to receive a relatively weak signal. The measure on how well a receiver receives is usually measured in millionths of a volt, commonly referred to as a microvolt.

A signal generator, calibrated in microvolts, (1/1,000,000 of a volt and millivolts 1/1000 of a volt) is the instrument used to identify how "sensitive" a receiver is. A signal generator is nothing more than a transmitter that produces relatively weak signals, just enough signal to check a receiver.

The frequency of the signal generator is set to the frequency the radio is receiving. Noise will be heard on the speaker when the signal generator is off and the radio receiver volume is turned up (the signal generator's output is connected directly to the radio's antenna jack).

With the volume on the radio still turned up, a signal from the generator is injected into the radio's antenna jack. As the generator signal is increased, the noise from the receiver will gradually decrease. The signal generator is calibrated on a dial to read, usually in one millionth of a volt increments, directly on the dial.

When a point is reached where the receiver is completely quieted as the signal generator level is increased, a final reading on the signal generator's dial is taken and recorded. The reading taken, usually in one millionth of a volt or microvolt on the dial of the signal generator, will be the voltage needed to completely quiet the receiver. This is what is known as "full quieting" a receiver.

4.   What is carrier power?

Carrier power is the actual power (expressed as true power in watts) developed in an antenna. True power is electrical energy dissipated in the form of heat in a pure resistive load (antenna or dummy load antenna). The effectiveness of alternating current is compared to Direct Current (D.C.). Alternating current has an effective value equal to direct current that is .707 times the alternating peak voltage (or current). This value is referred to as Root Mean Square (RMS) voltage or current. RMS value is the square root of the summed and averaged individually squared sine of each angle of the electrical unit being used (voltage or current) in one cycle of half time.

As an example, ten volts peak is the same as (10V) x .707 = 7.07 volts RMS. In other words, it takes ten volts of peak alternating current to equal 7.07 volts of direct current. Effective value of alternating current is also compared to heating value of direct current.

Stated another way, if it takes seven volts of direct current to brilliantly light a bulb, then it would take about ten volts of alternating current to get the same effect (7 volts DC / .707 RMS factor = 9.9 peak volts a.c. or about 10 peak volts of a.c.). Remember: alternating voltage or current is one peak positive (+) and one peak negative (-) shown in figure 35 below.

So, true carrier power is expressed in RMS values and is computed as follows: $P = V^2 \div R$. (P) is True Power, (V) is RMS Voltage and (R) is Resistance.

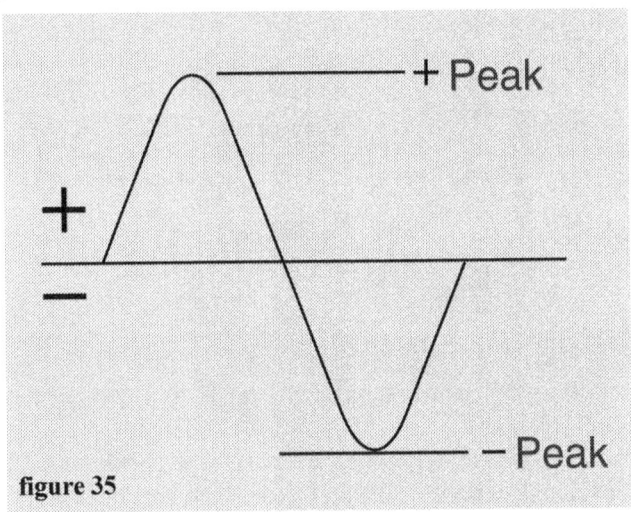

figure 35

Assume 13 volts peak. To drive a 50 Ohm antenna for transmitting, what would be the "true" or "RMS" power radiating out of the antenna? We know that in order to find "true" power, we must first convert the peak voltage reading to an "RMS" value. This conversion is made by following the example below. Note that voltage can be expressed by the letters "E" or "V" and used interchangeably.

Volts "RMS", is equal to (.707) x (V peak) = V_{RMS}. It follows then that (.707) x (13.0) = 9.19 volts "RMS".

Now by employing the "true" power equation:

Power = (Voltage^2 ÷ Antenna Resistance) = $P = V^2 ÷ R$.

We can now find the "true" power radiating from the transmitting antenna by substituting the 9.19 volts "RMS" into the power equation. So it would be:

$P = V^2 ÷ R$ where P = "true" power.

V = "RMS" voltage and R = Resistance of the radiating antenna. It follows then that $P = (9.19)^2 ÷ 50 = 1.7$ watts of "true" power.

So, power radiating from the antenna is computed to be approximately 1.7 watts. Practically speaking, checking power to an antenna is done simply by connecting a Wattmeter between the radio transmitter and the radiating antenna, as shown in figure 36 below.

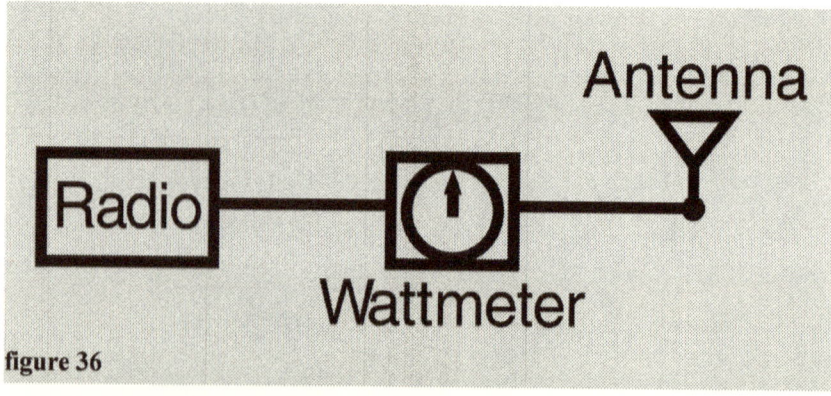

figure 36

5. What is Percentage of Modulation?

The variations on displacing the carrier voltage is said to be modulation. The amount of displacement of the carrier voltage will be determined by the audio signal voltage modulating the carrier. Expressed mathematically:

V_{MAX} = Maximum peak carrier voltage

V_{MIN} = Minimum peak carrier voltage

% Modulation = $V_{MAX} - V_{MIN}$ x 100 ÷ V_{MAX}

This displacement is represented in percentage of the carrier voltage. See figure 37 below:

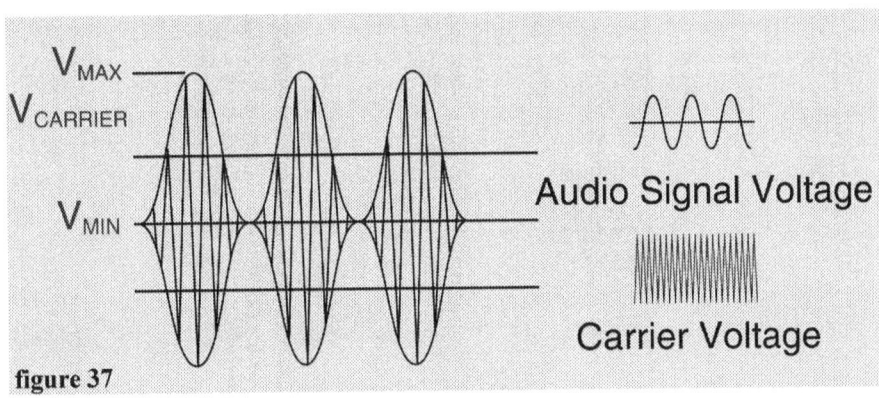

figure 37

6. How is modulation checked?

The best way to check modulation is with an oscilloscope. An oscilloscope is a testing device that displays a visual representation of a signal on a screen, similar to a television. Refer to figure 38 below:

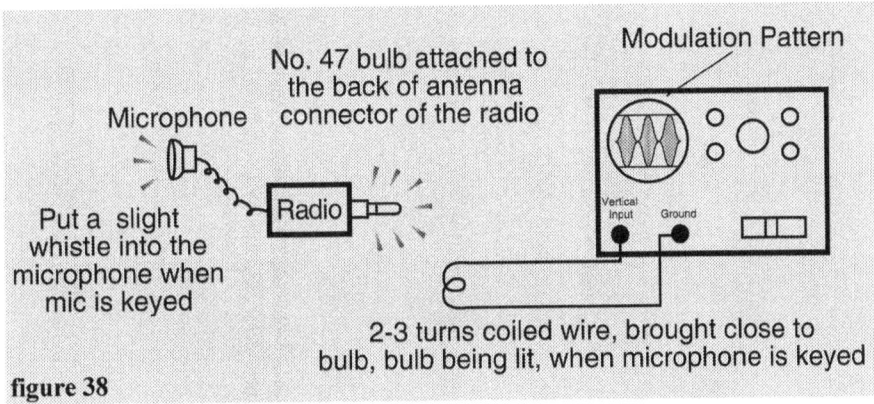

figure 38

7. What is over modulation?

A carrier is said to be 100% modulated if the audio modulating voltage equals twice the carrier voltage. See figure 39 on page 62:

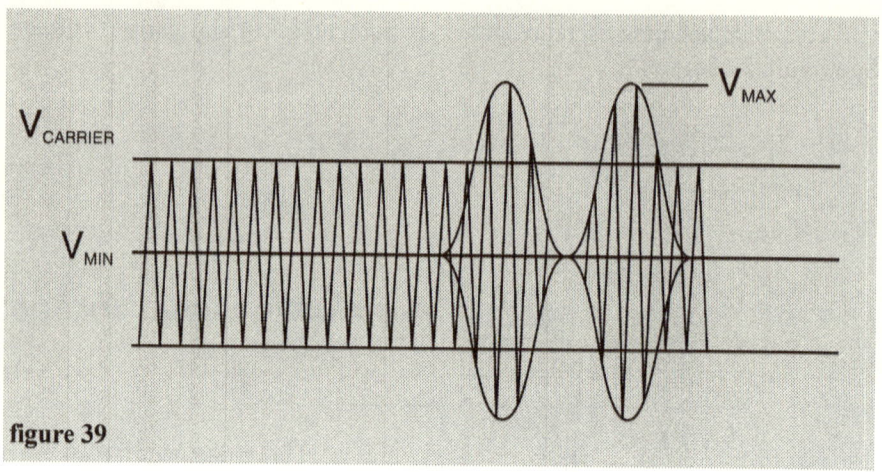

figure 39

A modulating voltage greater than twice the carrier voltage is over modulation. See figure 40 on the following page:

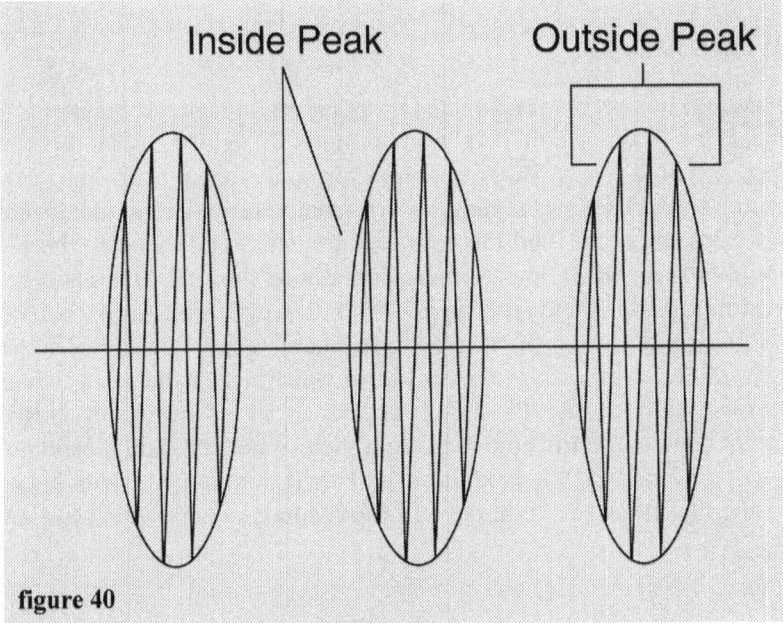

figure 40

Example: overmodulation (greater than 100%)

This results in spurious radio signals as well as audio distortion to whomever is receiving your transmitted signal.

Spurious radio signals can be harmonic radio frequencies. Harmonics are multiples of the original transmitting frequency. For example, the first harmonic frequency is the original carrier frequency. The second harmonic frequency is two times the original carrier frequency, the third harmonic frequency is three times the original carrier frequency, et cetera, et cetera. This is termed the order of harmonics.

Since the original carrier frequency is changed into harmonics from over modulation, 1st, 2nd, 3rd order harmonics, et cetera result in different frequencies being transmitted. This is what is referred to as "splatter". Since these harmonic frequencies are transmitted and are different from the original carrier frequency, they splatter over several bands of radio frequencies.

8. What is selectivity?

The ability of a receiver to select a channel (frequency) without having another nearby channel interfere with a selected channel is called selectivity.

9. What is peaking?

Peaking is the process of maximizing the magnetic lines of force produced by either a coil or a transformer. To peak a radio, first put a number 47 bulb on a PL-259 connector and attach it to the SO-239 connector on the back of the radio. Now, from your radio manual, locate the transmitter section and find the coil(s) or transformer(s). Locate the Ferrite core within the coil or transformer. Sometimes the coil or transformer is encased in a metal shield (see figure 41 for a sample coil form). Now, turn the radio on and key up the microphone, enabling the transmitter function. Next, turn the ferrite core in the coil form watching for the point where the number 47 bulb on the PL-259 connector is at its maximum brilliance during the time the ferrite core is being turned. When maximum brilliance of the number 47 bulb is achieved, lock the Ferrite core in place with bees wax or use the thin rubber band in the coil form, sometimes provided to lock the core.

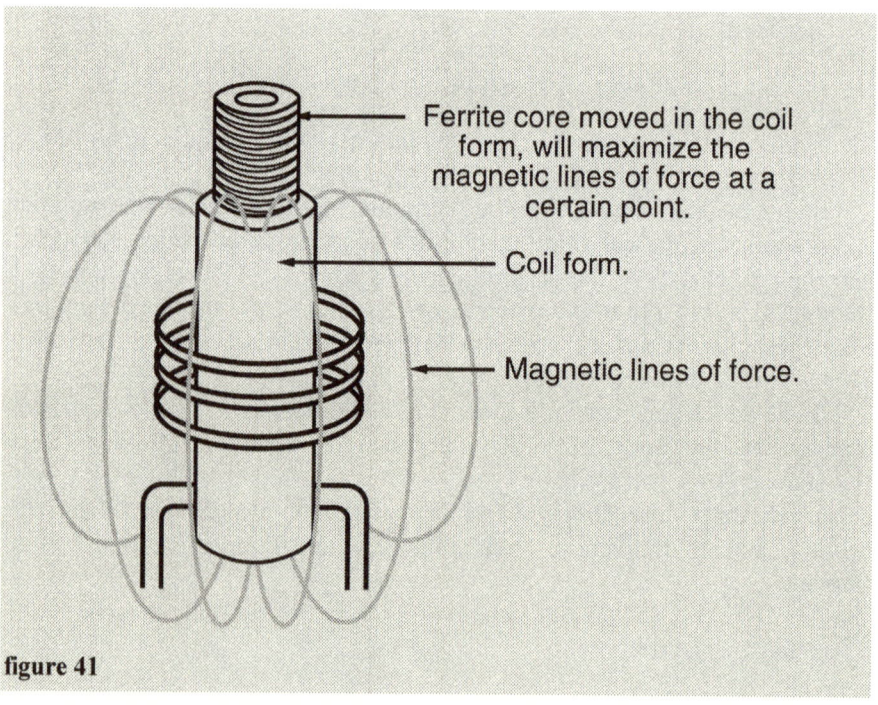

Ferrite core moved in the coil form, will maximize the magnetic lines of force at a certain point.

Coil form.

Magnetic lines of force.

figure 41

10. What size wire do I use for hooking up a C.B. radio?

Eighteen gauge wire is commonly used on C.B. radios. The lower the gauge, the thicker the wire. The longer the wire, the lower the gauge used. Wire exhibits a resistance. The longer the wire, the greater its resistance. Therefore, increasing the resistance also increases the voltage drop across the wire, leaving less usable voltage available at the radio (or device). A wire specification guide (available at most electrical or electronic supply companies) should be referred to for finding a specific wire size for the amount of current and the length of wire being installed.

You can also calculate the voltage drop of a wire knowing what the amount of current is going through the wire and what the resistance of the wire is per linear foot (see figure 42 for illustration of wire calculation).

a.   Eighteen gauge wire exhibits about .007 Ohm per foot.
b.   Current going through wire "A" is ten Amps.
c.   By Ohm's law formula, voltage = current x resistance.
    voltage = (10) x (.007) = .07 so voltage drop per foot is .07 volts.

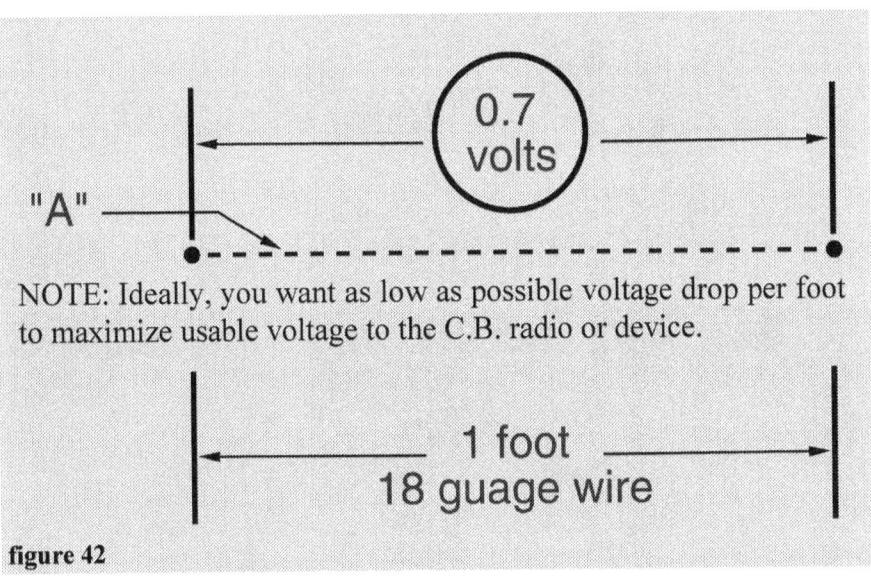

NOTE: Ideally, you want as low as possible voltage drop per foot to maximize usable voltage to the C.B. radio or device.

figure 42

11. What is a final amplifier?

A final amplifier is the last stage a carrier voltage goes through to be amplified, filtered and matched to the correct resistive load, that is the antenna (see figure 43 for identification purposes).

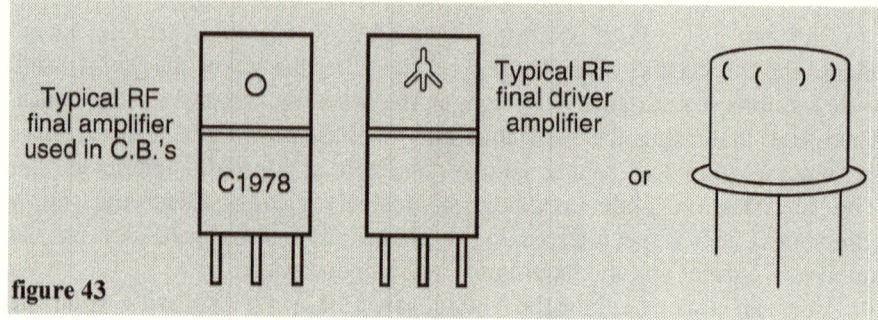

Typical RF final amplifier used in C.B.'s

C1978

Typical RF final driver amplifier

or

figure 43

12. What is an Integrated Circuit?

An integrated circuit, or I.C., consists of resistors, transistors and other components, miniaturized, grouped and manufactured in a plastic or ceramic package. Integrated circuits have different shapes, sizes and number of pins, as illustrated in figure 44 below.

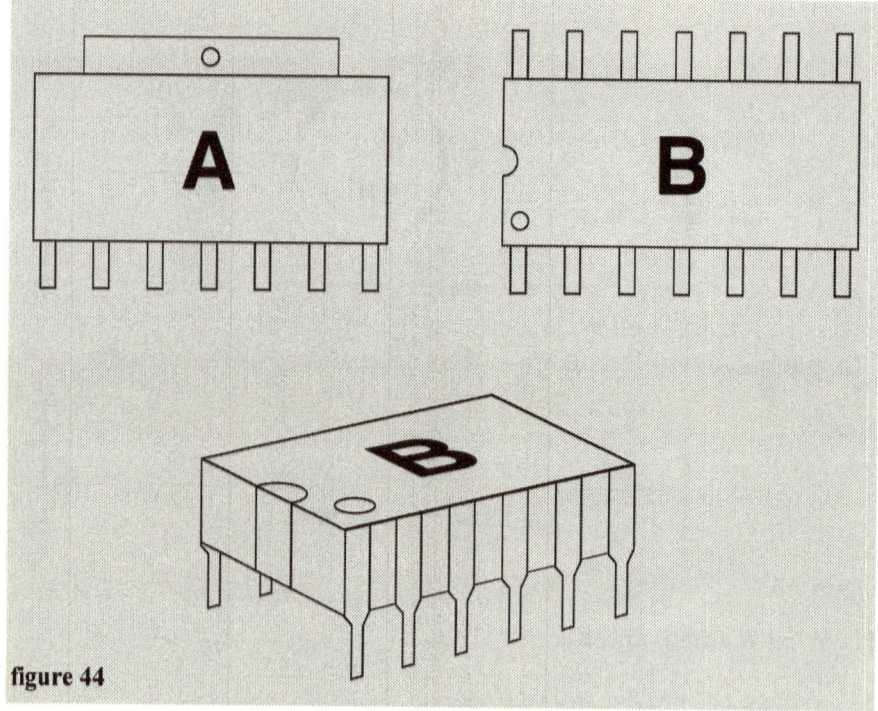

figure 44

13. What is a Power Microphone?

A power microphone is nothing more than a microphone with a built-in amplifier to help modulate the carrier. Since most radios have an internal adjustment for modulation and since you can only modulate a carrier to 100%, a power microphone has little value. If you are not getting enough modulation out of your radio, chances are, your power output is too high and you need to decrease the transmitter's R.F. output.

14. What kind of range can I expect from my radio?

The range of a C.B. radio transmitter varies with at least five things:

    a.   The placement and gain of an antenna.
    b.   The amount of loss in the coaxial cable from the radio to the antenna.
    c.   The amount of modulation added to the carrier.
    d.   The amount of power delivered to the antenna.
    e.   The amount of ambient and/or channel frequency noise present on the channel at the time of transmission.

As a general rule, one can expect, under ideal conditions (i.e. low noise and properly tuned radio to the antenna), about one mile per watt of power radiating from the antenna.

15. What is a Crystal?

A crystal is a piece of quartz crystal. If you were to take a crystal can apart, you would find a disc shaped piece of quartz with some metal electroplated on each side and a wire soldered to each plating. These wires are the leads to connect the crystal to the circuit. Quartz has the quality of reacting electrically to mechanical pressure by generating a voltage. If an alternating current is applied to a crystal, it will react vibrating at a frequency of the alternating current, provided the crystal is cut to vibrate at the same frequency as the alternating current frequency. For this reason, it is used in electronics. When used in an oscillator, by cutting a crystal to the proper thickness, it will maintain a specific frequency, thus controlling the oscillator frequency. The exact size, thickness and grain cut, will determine the resonant frequency the crystal will react to and it will hold that frequency.

16. What is true power and how does it relate to peak envelope power?

True power is expressed in RMS. True Power is .707 times peak current times .707 times peak voltage. True power applies when terminated in a pure resistive load and heat is dissipated.

Peak envelope power (PEP) is the instantaneous power transmitted when modulated 100% with an audio signal to its highest peak voltage V_{MAX} (see figure 39, page 62.) True power is in terms of carrier power, while peak envelope power is the modulation of the carrier power at a certain point in time. Peak envelope power formula is:

$$Pep = \frac{V^2_{RMS}}{R} \text{ or } Pep = V_{RMS} \times I_{RMS} \text{ or } Pep = I^2_{RMS} \times R$$

where Pep = Peak envelope power, V_{RMS} = Volts RMS of modulated carrier, I_{RMS} = Current RMS of modulated carrier and R = Antenna resistance

Remember, RMS values are used as effective values to equal the same heating effect of direct current.

Peak power is equal to four times the carrier power when modulated 100%. It is true, as peak voltage doubles, current doubles. As seen mathematically:

P = IV, where P = Power, I = Current and V = Voltage

So, if a carrier voltage is 10 volts modulated to 20 volts over a 50 watt (Ohm) antenna load, then it follows:

I unmodulated = 10 ÷ 50 = .2A. I modulated = 20 ÷ 50 = .4A, then P unmodulated = (.2)(10) = 2W and P modulated = (.4)(20) = 8 watts

17. What kind of power output can I expect from my radio and can it be peaked?

Typically, C.B. radios from the factory will have power output of about 4 watts. Adding a tone or voice to the carrier (modulated) will yield about 6 watts. This is what is known as Peak Envelope Power, or P.E.P. Most radios can be peaked for power output; however, there is a trade off at some point in the peaking process whereby the power on some radios is peaked so great that now there is very little modulating power left to properly modulate the carrier, so the audio on the carrier is significantly reduced.

18. What is Dummy Load and how is it used?

A dummy load is a resistance attached to the output of the transmitter to prevent unwanted radiation from the transmitter. A dummy load simulates an antenna and is in the form of a large (power) resistor. Dummy loads are rated for different power levels depending on the output power of the transmitter being tested. Dummy loads are rated on how much power (heat) is needed to be dissipated.

19. What is R.F. drive and how is it used?

R.F. drive usually refers to the signal needed to drive the final power amplifier on the transmitter sufficient enough to provide nominal power to the antenna (R.F. refers to radio frequency). R.F. is any signal close to 30 KHz (Kilo Hertz or cycles per second) and greater. In this case, the signal is a wave produced electrically, that shapes and repeats itself 30 thousand times every second. This wave produces a varying voltage, both positive and negative. This varying voltage is often referred to as Alternating Current or A.C.

20. What is the R.F. gain knob for?

The R.F. gain knob (or switch) is used to increase the sensitivity of the C.B. receiver. Increasing the gain increases the sensitivity of the receiver, increasing the receivers' ability to pick up weak signals.

21. What is the A.N.L. knob for?

A.N.L., meaning automatic noise limiter, is used to reduce inherent low frequency noise in the receiver circuit.
There are feedback circuits in the receiver circuit, whereby, if the receiver receives an intense noise signal, the feedback circuit will operate to reduce the incoming noise signal as it gets increasingly stronger. So when the noise signal increases, the feedback circuit signal also increases, but is phased to counter the incoming noise signal and reduce the noise on the receiver output.

22. What is the relationship of carrier power to modulating power?

As you increase your carrier power, you must also increase the audio power as well. If this is not done, you will have a substantially weaker audio signal on the carrier. Here is how it looks from a mathematical model:

$P_{AF} = M^2 P_{DC} \div 2$, where P_{AF} = audio frequency power needed
M = modulation % expressed as a decimal (i.e. 0.2 = 20%)
P_{DC} = direct current carrier input power to the final power amplifier

For example, to modulate a final power amplifier 100% when its D.C. input is 500 watts, then:

$P_{AF} = M^2 P_{DC} \div 2 = (1^2)(500) \div 2 = 250$ watts

For the sake of illustration, let's increase the final amplifier input power from 500 to 1, 000 watts. However, we will use the 250 watts of modulating power used for a 500 watt final amplifier. The % of modulation is:

$$M = \sqrt{2P_{AF} \div P_{DC}} = \sqrt{2(250) \div 1,000} = 0.7\%$$

In this example, notice as the carrier power went up, the % of modulation went down.

23. How do I replace the PL-259 UHF male connector?

Figure 45 provides an illustration to follow in replacing a PL-259 UHF male connector.

a. First, please note the coaxial cable in figures 45 and 46 on page 71. There are four parts to this cable: an outer insulator, a braided wire, a center conductor wire, which is inside the fourth part, and an inner insulator.

b. Measure about 1½ inches from the end of the coaxial cable. Using a sharp knife, cut the outer insulation, being careful not to cut the braided wire cable inside, and remove it. Now slide the threaded sleeve insert over the end of the coaxial cable putting the threaded end of the sleeve close to the point where the outer insulator of the coaxial cable was cut and removed.

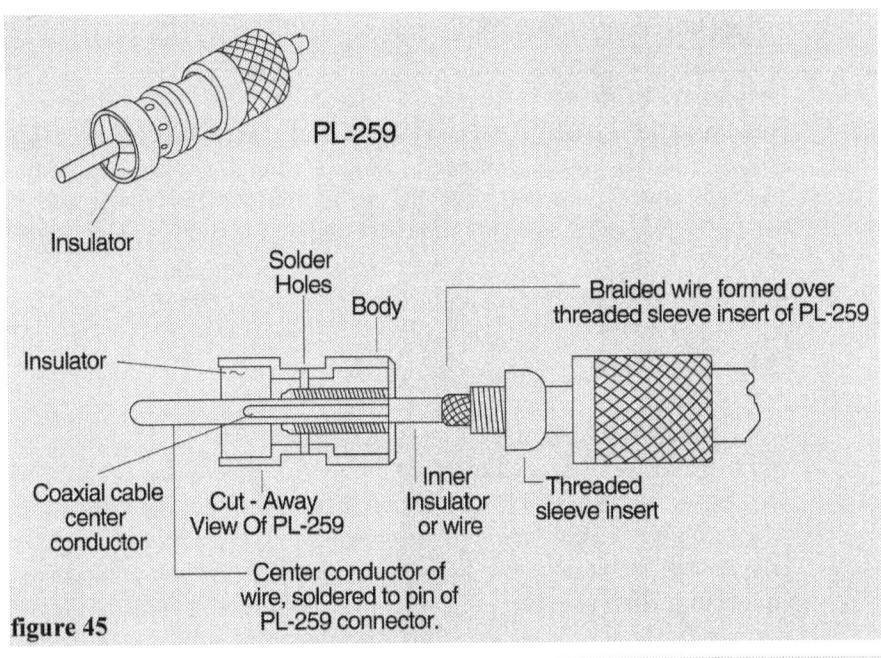

figure 45

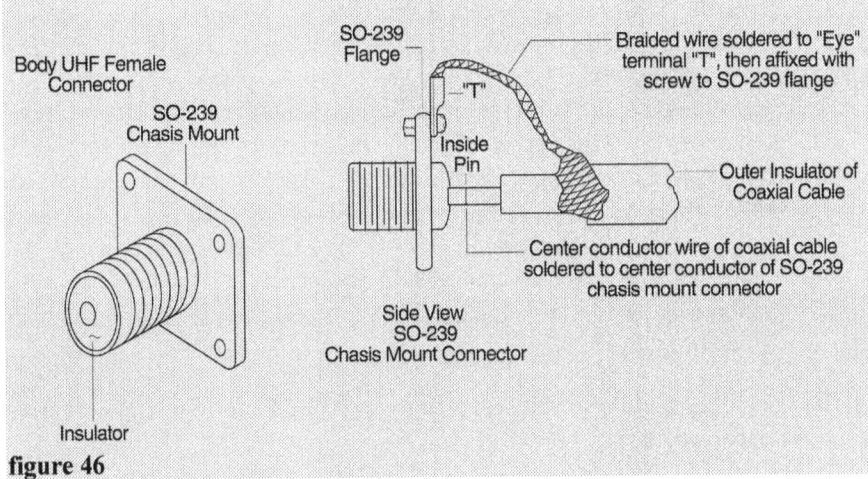

figure 46

c. Form the braided wire over the portion on the sleeve insert closest to the threads.

d. Remove any excess braided wire so that it does not get caught in the threaded sleeve insert threads.

e. With a sharp knife, cut the inner insulator about ½ of an inch outward from where the last cut was made on the outer insulator of the coaxial cable and remove it from the wire. Exercise care when cutting the inner insulator to prevent cutting into the center wire conductor. The center conductor wire should have about 1½ inch length wire exposed.

f.   Now take the PL-259 UHF connector and insert the coaxial cable with the sleeve insert tightened. Make sure the center wire of the coaxial cable is inserted into the pin of the PL-259 connector. If excess wire comes through the pin, remove it with wire cutters. Wire should be cut flush with the center conductor pin.

g.   Rotate the PL-259 connector until you can see the braided wire from any one of the five holes around the PL-259 connector.

h.   Using a 200 watt solder iron, solder one of the PL-259 holes to the braided wire.

i.   Next, attach one lead of the Ohmmeter to the body of the connector and the other lead to the center conductor of the coaxial cable to be sure there is no short. If it is shorted, the reading on the Ohmmeter will be about .4 Ohm and you will need to cut the connector off the coaxial cable and unsolder the cut piece of coaxial cable inside the connector. Then resolder the connector. Then repeat steps f-i retest it to verify there is no short and the Ohmmeter reads in the millions.

j.   With the Ohmmeter disconnected, the next step is to solder the center wire of the coaxial cable to the center pin of the PL-259 connector.

k.   Reconnect the Ohmmeter attaching one lead to the body of the connector and the other lead to the center pin of the PL-259 connector. If this connection is shorted once again, the reading on the Ohmmeter will be about .4 Ohm. Should this be the case, check for the center conductor pin wire being shorted to the body of the connector. Unsolder the connection and repeat steps f-i. Retest with the Ohmmeter to verify there is no short and the Ohmmeter reads in the millions.

24. How is a coaxial cable installed onto an SO-239 female UHF chassis mount connector?

The procedure for installing and testing a coaxial cable onto an SO-239 female UHF chassis mount connector, shown in figure 46 on page 71, is as follows:

a.   Measure back about ½ inch from the end of a coaxial cable. Using a sharp knife, cut the outer insulator around the diameter of the wire. Use caution when cutting through the outer insulator so you don't cut into the braided wire. Slide the ½ inch cut piece of outer insulator off the coaxial cable. The braided wire is now exposed.

b. Separate a portion of the braided wire with a pointed object. The separated portion of the braided wire now exposes the center wire insulator.

c. Bend the center wire and inner insulator 180° parallel to the coaxial cable.

d. Using an awl or other pointed object, pull the center wire and insulator through the braided wire.

e. Strip about ⅛ of an inch off the inner insulator measuring from the tip back.

f. Now, using a 100 watt soldering iron, solder the center wire to the inside pin of the SO-239 connector.

g. Check for a short of the inside pin to the outside body of the connector by placing one lead of an Ohmmeter on the resistance times one (R x 1) scale to the inside pin of the SO-239 connector and the other lead to the body of the connector. The Ohmmeter should read several million Ohms of resistance. Should you not get this high reading, you have a short and will need to remove the soldered coaxial cable lead from the connector and repeat the instructions, beginning with a.

25. How is the coaxial cable installed on an RCA connector?

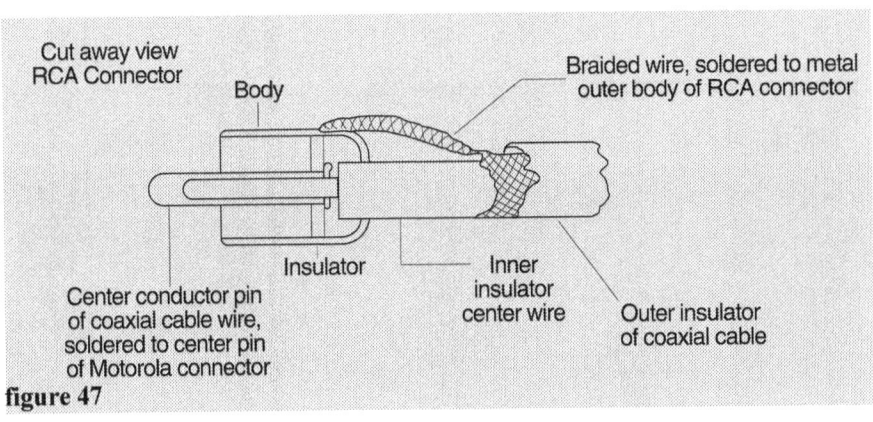

figure 47

Figure 47, above, provides an illustration of an RCA connector.

a. Measure approximately 1½ inches back from one end of the coaxial cable and cut the outer insulator completely around the cable. Again, use caution and don't cut into the braided wire.
b. Remove the 1½ inch cut piece of coaxial insulator so the braided wire is exposed.
c. Now spread the braided wire and expose the coaxial cable inner insulator and center wire at one point on the center wire.
d. Bend the center wire and inner insulator 180 degrees at the point where the braided wire was spread. Using an awl, insert it at the point where the braided wire was spread and pull the center wire and insulator through the braided wire.
e. Measure back from the tip of the center wire about 1 inch. Using a sharp knife, cut the insulator at the 1 inch mark. Use caution and don't cut the center wire conductor.
f. Slide the 1 inch piece of insulation off the center wire and discard.
g. Insert the center wire conductor into the center pin of the RCA connector until the inner insulator stops the wire from going further.
h. Cut any excess wire off the tip of the center pin of the RCA connector.
i. Now, using a 100 watt soldering iron, solder the center wire to the center pin and braided wire to the outside body of the connector.
j. Use the same procedure with the Ohmmeter as instructed in question 23, instruction i, to verify no short exists.

26. How is coaxial cable installed on a Motorola connector?

Figure 48, below, provides an illustration of a Motorola connector.

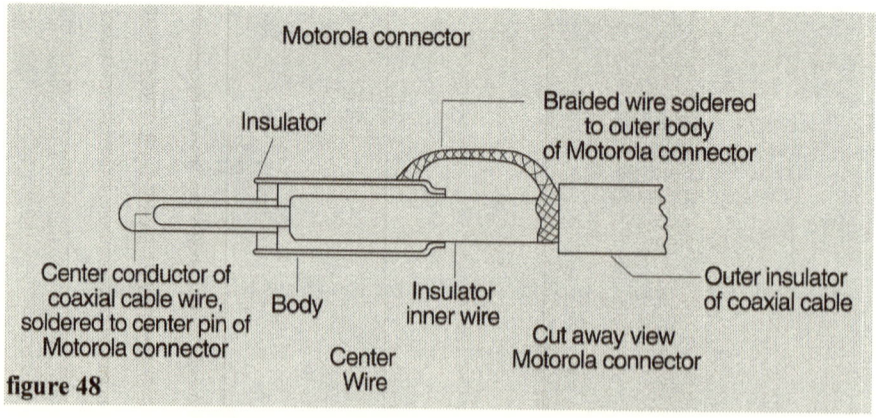

figure 48

The procedure for installing and testing a coaxial cable on a Motorola connector is as follows:

a. Measure approximately 1½ inches back from one end of the coaxial cable and cut the outer insulator completely around the cable.
Again, use caution and do not cut into the braided wire.

b. Remove the 1½ inch coaxial insulator cut so the braided wire is exposed.

c. Now spread the braided wire and expose the coaxial cable inner insulator and center wire at one point on the center wire.

d. Bend the center wire and inner insulator 180 degrees at the point where the braided wire was spread. Using an awl, insert it at the point where the braided wire was spread and pull the center wire and insulator through the braided wire.

e. Measure back from the tip of the center wire about 1 inch. Using a sharp knife, cut the insulator at the 1 inch mark. Use caution and don't cut the center wire conductor.

f. Slide the 1 inch piece of insulation off the center wire and discard.

g. Using a 100 watt soldering iron, solder the inner center wire to the center pin of the connector.

h. Use the same procedure outlined in i of question 23 to detect a short.

27. What is VSWR?

If you have been involved with C.B. radios for any length of time, no doubt you are familiar with Voltage Standing Wave Ratio or VSWR.

Standing waves are reflected components of either voltage or current from the resistive load you are driving (the antenna). If the load you are driving is an antenna from a C.B. transmitter, and it is not a proper resistive match, a reflective signal will result back into the transmitter.

The reflected signal is referred to as the reflected wave from the load. The driving signal to the antenna (load) is what is referred to as the incident wave from the source. See figure 49 on the following page:

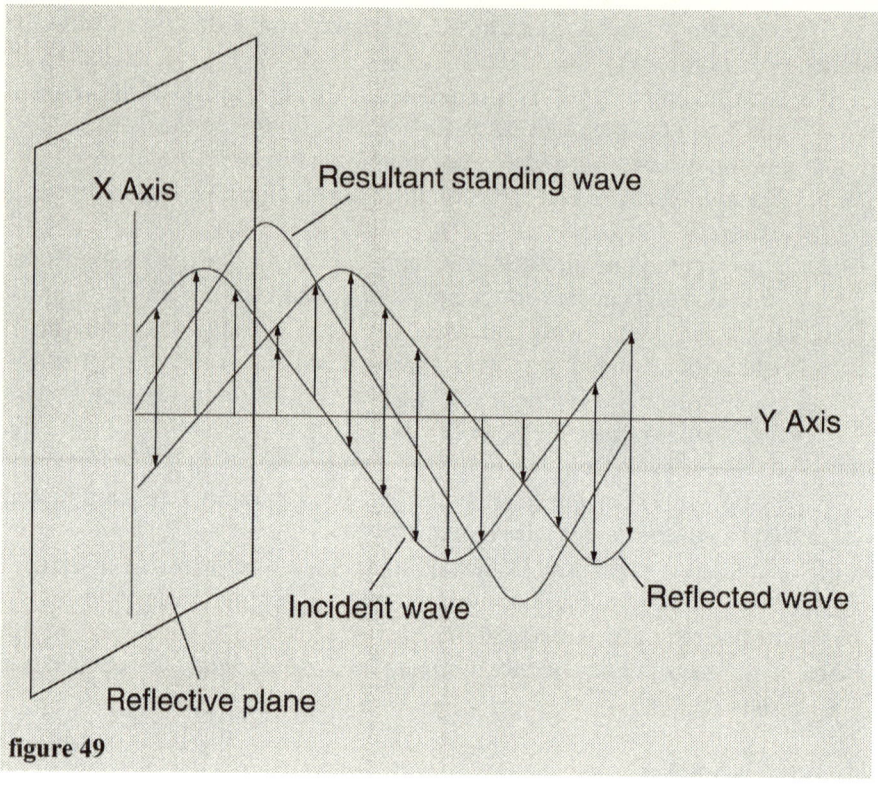

figure 49

We will use a light source to illustrate reflected waves. When you hold a flashlight to a mirror or shiny object, the light is not absorbed but rather, reflected back. But if you hold a flashlight to a dark object, the light energy is absorbed rather than reflected back.

The reflected light theory is analogous to the electric wave theory.

If you transmit electrical energy to a properly tuned antenna with the correct resistance, this medium will absorb all this electrical energy. The amount of driving energy compared to and combined with the amount of energy being reflected is referred to as standing waves.

Another example of standing waves can be expressed in the transient movement of the strings of an instrument which are plucked, such as the guitar. The strings vibrating with an up and down motion on a stationary plane is an example of what a standing wave is. The same thing occurs for an electrical wave. See figure 50 on page 77:

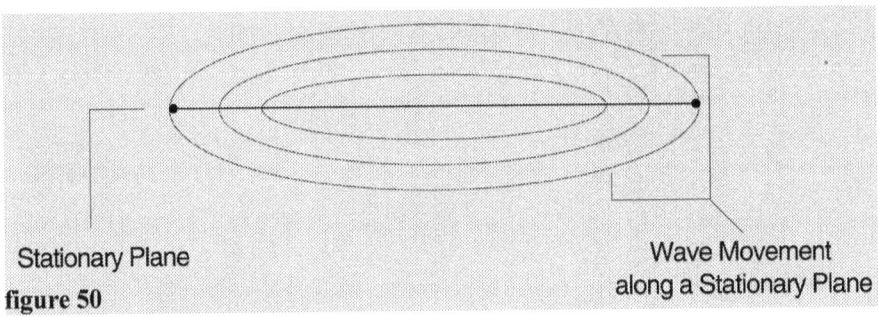

Stationary Plane

Wave Movement
along a Stationary Plane

figure 50

28. What is Harmonic Distortion?

Harmonic frequencies are multiple frequencies of the original frequency. They may be 1st, 2nd, 3rd...order harmonics, i.e. (1) original frequency, (2) 2 times original frequency, (3) 3 times original frequency, et cetera. They result from audio frequency distortion in the microphone circuit. Harmonic distortion is a result of over modulation, when modulated with a carrier frequency, whereby the original audio signal, when over driven, no longer is a faithful reproduction of the original input signal. See figure 51 below for an example of harmonic distortion as seen on an oscilloscope.

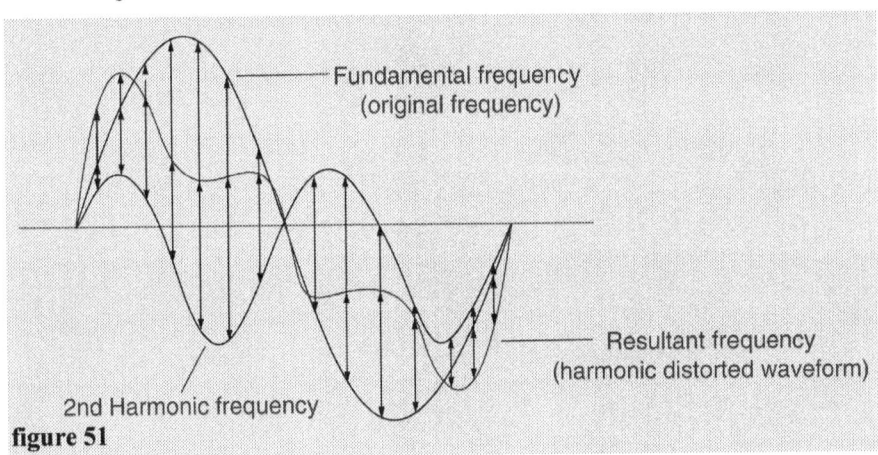

Fundamental frequency
(original frequency)

Resultant frequency
(harmonic distorted waveform)

2nd Harmonic frequency

figure 51

To reduce the possibility of harmonic distortion, reduce the microphone output audio signal on the radio to the point where the output pattern, as seen on the oscilloscope, resembles figure 52 below:

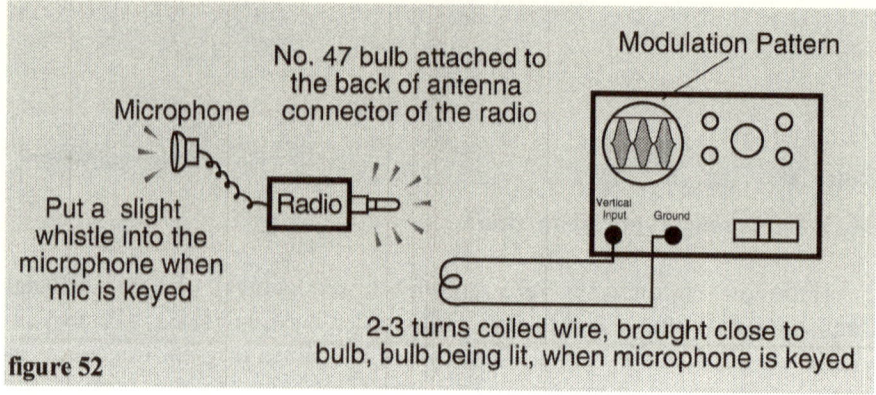

No. 47 bulb attached to the back of antenna connector of the radio

Modulation Pattern

Microphone

Put a slight whistle into the microphone when mic is keyed

Radio

2-3 turns coiled wire, brought close to bulb, bulb being lit, when microphone is keyed

figure 52

29. What is heatsink compound?

Heatsink compound is a white pasty or greasy translucent material used to help transfer heat from an electronic device to a frame of a radio or any other metallic object for heat dissipation. For the best conduction of heat transfer, the better one to use is the white pasty heatsink compound.

A POLARITY TESTER

In order to prevent reversing polarity on your radio and causing costly damage, there is a very handy circuit you can build for a few dollars, used to determine the polarity (+ or -) of power leads on a vehicle or power supply used to power a C.B. radio or other electronic devices. (See figure 53 for a visual look at the tester before starting to purchase parts and assemble.)

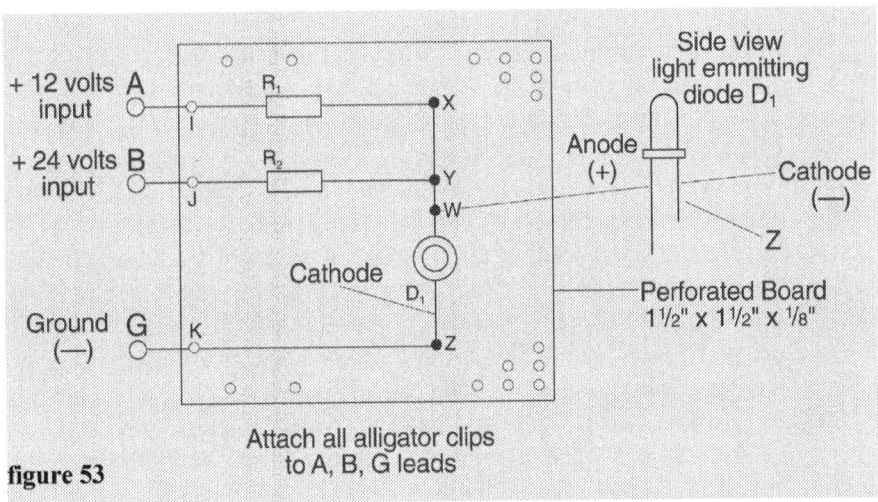

figure 53

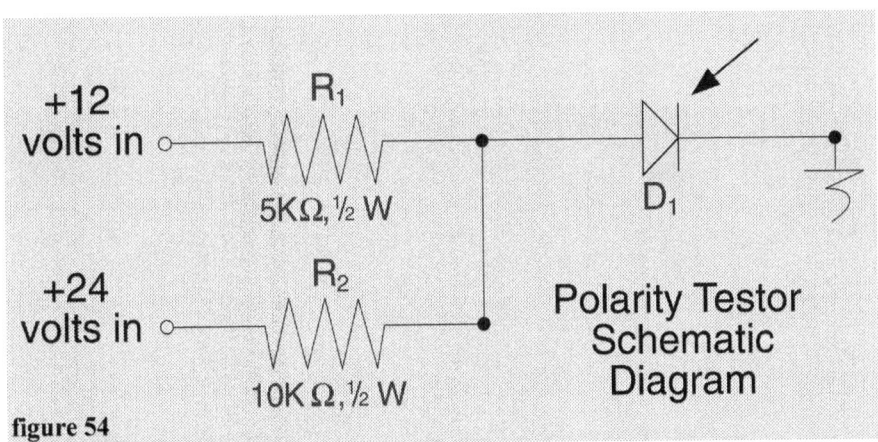

figure 54

79

Polarity Tester Assembly Instructions

1. Purchase parts on parts list.
2. Insert both ends of R_1 resistor into any two holes in the perforated board illustrated in figure 53.
3. Now insert both ends of R_2 resistor into any two holes in the perforated board illustrated in figure 53.
4. Connect and solder points "X" and "Y" of resistors R_1 and R_2 respectively shown in figure 53.
5. Insert light emitting diode (D_1) into any two available holes on the perforated board with the anode (longest lead of the light emitting diode) lead positioned closest to point "Y" located on R_2 shown in figure 53.
6. Connect and solder the Anode lead of D_1 point (light emitting diode) "W" to point Y of R_2 shown on figure 53.
7. Locate the short lead of D_1. This is the cathode lead marked "Z". Insert and solder one end of a piece of 20 gauge wire to point "Z" of D_1. Route and connect the other end of this wire through any hole of the perforated board. This will be labeled point "G" and will go to the negative ground shown on figure 53.
8. Now solder two 12 inch lengths of stranded 20 gauge wire to points "I" and "J" of R_1 and R_2, shown on figure 53. Affix an alligator clip to the other end of these 20 gauge wires.
9. Now solder the third 12 inch length of stranded 20 gauge wire to point "K" as shown in figure 53. Affix an alligator clip to the other end of this 20 gauge wire.

At this point, the polarity tester circuit is assembled and ready to use.

a. Take the ground (G) lead on the polarity tester, attach it to the vehicle ground or power supply ground if used **(insure that no power is applied to the circuit under test at this time)**.

b. **With the power still off**, select either the "A" or "B" lead from the polarity tester shown on figure 53 and attach it to the positive (+) source of the circuit under test.

c. Now apply power. If the polarity tester fails to illuminate, switch the ground (G) lead on the tester to the other terminal under test. The polarity tester light (light emitting diode, LED) should light. If the test (LED) illuminates brightly, **remove the power** and then switch the "A" and "B" leads.

Parts List For Polarity Tester

Part Description	Quantity	Schematic Description	Electronic Supply Companies
Resistor, 5K Ohm, ½ watt	1 ea.	R_1	by component value
Resistor, 10K Ohm, ½ watt	1 ea.	R_2	by component value
Light Emitting Diode LED (red)	1 ea.	D_1	1 to 3 milliamp 1.8 volts, D.C.
Perforated Board, 1 ½" x 1 ½" x ⅛"	1 ea.		By description 1 ½" x 1 ½" x ⅛"
Alligator clips	3 ea.		By description
12 inch, 20 gauge stranded wire	3 ea.		By description

D.C. – D.C. CONVERTER

Have you ever wished you had an electronic device that could convert your vehicle's electrical system from 24 volts to 13 volts in order to operate a CB radio or similar electronic devices?

There is an electronic device that does exactly that: converts one D.C. voltage to another.

This sort of device is called a DC-to-DC converter. One is illustrated below.

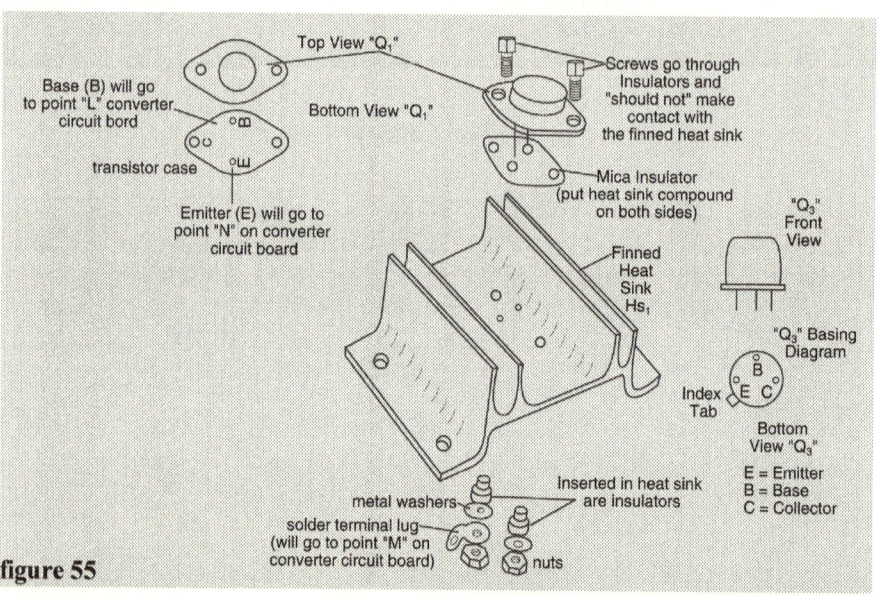

figure 55

This device will change 24 volts DC (direct current) to approximately 13.6 volts DC, with a maximum current draw of about 6 amps for 30-second intervals, and a draw of 3 amps under continuous operation.

The converter circuit is relatively straightforward and easy to build. If you shop around at local electronics parts supply stores, you can probably acquire everything you need to build this converter for $5 or less.

INSTRUCTIONS FOR ASSEMBLY

1.   Begin by referring to figure 56 for circuit board assembly of this DC-to-DC converter.

2.   Locate transistors Q_2 and Q_3 on perforated board in Figure 56. Be sure to make note of the basing diagram of Q_1, Q_2 and Q_3. (Figure 55, 56 respectively) This diagram will show you the connections for each of the three transistors.

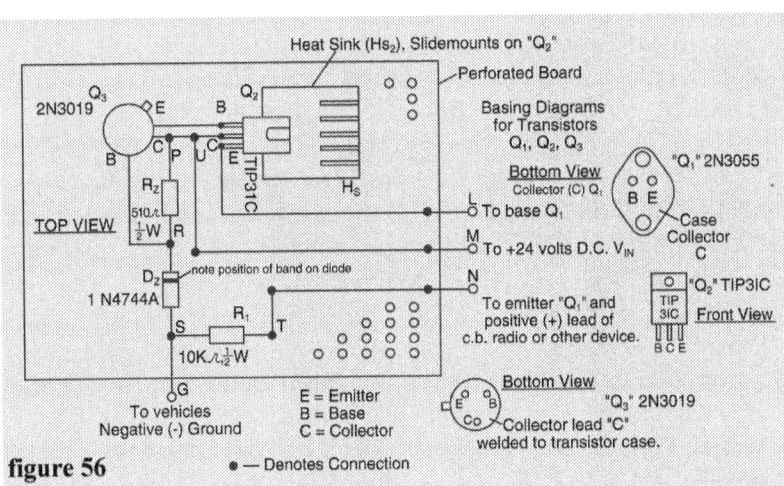

Parts Layout D.C. Converter Circuit Board

figure 56   • — Denotes Connection

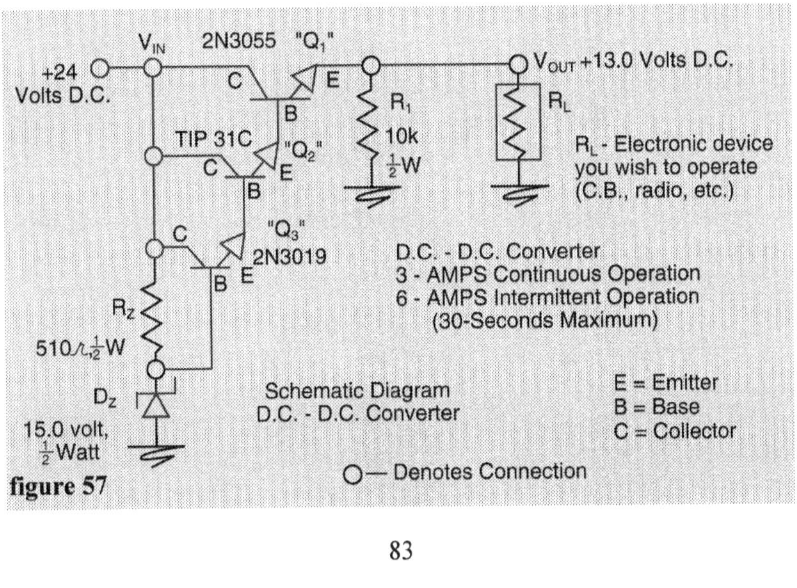

figure 57

3. Begin by inserting the three leads from Q_3 (2N3019) into the perforated board. Find Q_2 (TIP31C) on the basing diagram, located on the same page as Figure 56. Insert the leads from Q_2 into the perforated board as well. NOTE: Be sure to insert and position Q_2 and Q_3 leads into the perforated board EXACTLY as shown on the basing diagram.

4. With the leads of Q_2 and Q_3 inserted into the perforated board, connect the base lead of Q_2 to the emitter lead of Q_3. Solder these two connections together by using a 40-watt soldering iron.

5. Locate the collector leads ("C") on each of Q_2 and Q_3. Connect and solder the "C" of Q_2 onto the "C" of Q_3. Locate the zener diode resistor (R_Z) on the component layout. Locate point "P" on R_Z. Insert the resistor into the perforated board. Connect and solder one connection of R_Z to the "C" connection of Q_2, and the other to the "C" connection of Q_3.

6. Locate the zener diode (D_Z) on figure 56. Insert this diode into the perforated board. Connect and solder the lead closest to the "diode band" on D_Z directly to the other end of R_Z, at the point marked "R". Solder a short piece of 22ga solid wire at this same point "R", soldering the other end of the wire to the base (B) of Q_3.

7. Locate resistor R_1 on figure 56. Insert both leads of R_1 into the perforated board. Connect and solder one end of R_1 to the side of D_Z marked "S". Also at this point "S", solder a piece of 18ga stranded wire that is approximately 8 to 12 inches long. This wire will be connected to vehicle negative ground, marked "G".

8. Locate point "T" of R_1 as shown on Figure 56. Connect and solder one end of a piece of 20ga solid wire to this point "T". Insert the other end of this wire into the perforated board at the point marked "N".

9. Locate lead "M" on the converter circuit board on figure 56. Insert both ends of a piece of 20ga wire into the perforated board. Connect one end marked "U" to the collector leads ("C") of Q_2 and Q_3 that are already tied together. Once connected to "U", solder in place. The other end of this wire should be marked as point "M". This lead will be connected to the plus-24-volt circuit of the vehicle, marked +24 Volts D.C. Vin.

10. Locate point "L" on the converter circuit board. Insert both ends of a piece of 20ga wire into the perforated board. Connect and solder one end

of this wire to the emitter (E) of Q_2. The other end of this wire is marked "L", which goes to the base (B) of Q_1, as shown on figure 55.

11. Locate the heat sink (H_{S2}) on Figure 56. This type of heat sink will simply slide onto the transistor Q_2.

At this point, construction of the DC-to-DC circuit board is complete. What follows are the steps necessary for mounting Q_1 onto its heat sink.

12. Refer to transistor Q_1 and the finned heat sink as shown on figure 55. In addition to these two items, an installation kit is also required. These parts - named on the attached materials list - include screws, mica insulator, additional insulators, solder terminal lug, metal washers, plus two insulators that are fitted into the heat sink.

13. Insert the two screws from the mounting kit into the mounting case of transistor Q_1 (2N3055). Apply heat sink compound to both sides of the mica insulator, taking care to spread the compound liberally with your finger. Place the mica insulator onto transistor Q_1, noting that the insulator will only fit in one direction with the mounting holes of each item in alignment. This placement is clearly illustrated on Figure 55.

14. Set transistor Q_1 on the heat sink, as shown on Figure 55. Slide the two insulators over the Q_1 mounting screws. Place the two metal washers from the mounting kit over these screws, and follow by adding the solder terminal lug to either one of these screws.

15. Finally, thread the two fastening nuts from the mounting kit onto the mounting screws, using a screwdriver to hand-tighten them. Take care not to tighten to the point where the plastic insulators could be collapsed from the pressure.

16. To ensure that the Q_1 transistor has been properly installed onto the heat sink, use an ohmmeter and select the Rx1 scale. Place one of the ohmmeter's leads against the transistor case, and the other lead against the heat sink. Hold both leads firmly to ensure a good electrical connection. A resistance reading into the meg-ohm range indicates a good installation. A resistance reading of one ohm or less indicates a short circuit between the transistor and the heat sink. In this event, retrace your steps to make sure that the various insulating items are installed correctly, so that the mounting hardware is not shorted out against the heat sink.

Once the Q_1 transistor is properly mounted onto the heat sink, it is now time to connect the leads from the converter circuit board (figure 56) to those of transistor Q_1 (figure 55).

17. Locate point "L" on the right side of the converter circuit board (figure 56). Solder one end of 18ga stranded wire, approximately six inches long, to point "L", and the other end to base (B) of transistor Q_1.

18. Locate point "M" on the converter circuit board, and solder two pieces of 18ga stranded wire, each approximately six inches long, to this point. The end of one of these wires is soldered to the collector (C) lead on transistor Q_1. The end of the other wire is soldered to the +24 VDC lead from the vehicle. It is strongly advised that you place a 7-amp fuse in series with this latter wire.

19. Lastly, locate point "N" on the converter circuit board. Solder two pieces of 18ga stranded wire, each approximately six inches long, to this point. The end of one of these wires is soldered to the emitter (E) on transistor Q_1. The end of the other wire is soldered to the positive (+) lead of the electronic device that is being installed in the vehicle.

At this point, construction of the DC-to-DC converter is complete. Now it is time to test the circuit.

Install the converter circuit in the vehicle. As noted on figure 56, point "M" attaches to the vehicle's +24v power source, while point "G" is attached to the vehicle's ground. For testing purposes, DO NOT connect your electronic device (radio, etc.) to this circuit. Use a DC voltmeter at a range of 15 volts or higher, reading the output voltage at converter point "N" and vehicle ground. You should see a reading of around 13.6v DC. If the reading varies from this mark by 0.3v or greater (i.e., higher than 13.9v or lower than 13.3v) - or if there is a reading of zero volts - do not use this converter until the problem has been corrected.

A qualified electronics technician should be consulted to troubleshoot this converter circuit if it fails to function properly.

Parts List for the DC-to-DC Converter

One each of the following items:

DESCRIPTION	MFG. PART NO.
Transistor Q_1	2N3055
Transistor Q_2	TIP31C
Transistor Q_3	2N3019
Zener diode D_Z	1N4744A
Zener resistor R_Z	510 ohm, 1/2 watt
Resistor R_1	10K ohm, 1/2 watt
Heat sink H_{S2} for Q_2	Heat Sink for T0-220 transistor case
Perforated board	EXP-300PC (5x5 inches by 1/8 inch)
Heat sink H_{S1} for Q_1	Finned Heat Sink for T0-3 transistor case
Heat sink paste	Tech Spray or other brand
60/40 solder	18ga diameter

Also required:
40-watt solder iron

Please note that most electronic supply houses will stock these items by item or the manufacturer's parts number.

Sheldon A. Chrysler

Glossary

Alternating Current (a.c.):

Current that is continuously changing polarity from positive to negative and negative to positive. The amount of times alternating current changes polarity per unit of time varies directly with the frequency of the current.

Amplifier:

A device used to intensify a weak signal. A weak input signal going through an amplifier produces a stronger output signal.

Amplitude Modulation (A.M.):

A continuous waveform that varies in magnitude when a varying waveform is superimposed on it.

Antenna Electrical Length:

The computed half wave length of an antenna at a given operating frequency. The formula for a half wave length antenna is:

Half wave length (λ/ 2in feet = 492÷F$_{(MHz)}$
Example: λ/2= 492 ft/27 MHz = 18.2 feet
where 27 MHz = operating frequency of 27 Mega Hertz for C.B. radio.

The actual physical length of a half wave length antenna takes into account a condition called "end effect". End effect makes the electrically computed length of an antenna appear longer than the actual physical length of the antenna due to conductivity of air. Actual length of an antenna is about five percent of the electrical length. So the actual length of a half wave length antenna is

(.05) x 492 ft = 468 therefore the actual wave length (λ /2) of an antenna is λ/2= 468/F$_{(MHz)}$

Audio:

Audio is considered to be a tone, voice or sound heard by the human ear. An audio signal (sound) will be anywhere from about fifty to eighteen thousand cycles per second.

Carrier:
An electronic mobile means of travel for voice, data and other transmission of signals.

Coaxial Cable:
A cable which has a conductor parallel to another conductor with an insulator in between both conductors.

Co-phase:
Using a common phase of a radio signal as in the case of co-phased (dual) antennas, one phase of the signal will go to one antenna, while the other phase will go to another antenna.

Current:
The quantity of electron flow through a conductor per unit of time.

Cycle:
An event that occurs and repeats itself per unit of time. In terms of electronics, it is considered to be the amount of time an electrical waveform shapes and repeats itself over a period of time. An example is one-half rotation of a circle is one-half cycle.

Diode:
An electronic device that will allow the flow of current in only one direction when the proper voltage polarity (+ or -) is applied.

Dipole Antenna:
A dipole has two horizontal radiating elements constantly changing polarity with a change of time and radiation pattern produced in a plane perpendicular to its axis. The radiating elements are center fed by the transmission line.

Direct Current (D.C.)
Electrical current that flows in one direction at a rate that varies inversely with the load and directly with the voltage.

Dummy Load:
A resistance that simulates the resistance of an antenna. It does not radiate the energy as an antenna, but rather, absorbs it. It is sometimes referred to as an artificial antenna.

Electronic Power Supply Filter:

A network of resistors, coils and capacitors used to suppress unwanted signals and electrical noise, like noise from spark plugs, the alternator, et cetera.

Effective Radiated Power (ERP):

Carrier power transmitted by an antenna as a result of power gain of the antenna. Mathematically, P_{ER} = effective radiated power, P_{IN} = radio frequency carrier power to the input of an antenna, G_P = power gain of the antenna (power is expressed in watts). $P_{ER} = P_{IN} \times G_P$

Electromagnetic:

A magnetic field set up perpendicular to a conductor that is created as a result of current (electron) flow through a conductor.

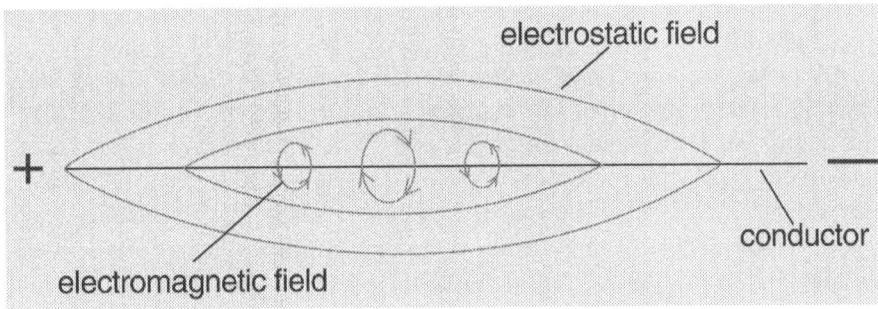

Electrostatic:

A stationary charge of electromotive force. That is the accumulation of electrons not in motion. Positively charged and negatively charged particles not in motion. In order to get this stationary negative charge in motion, it will be necessary to have both or either a negatively charged source to repel or positive charged source to attract an electron to provide movement and current flow.

Ferrite Core:

A compressed powered iron material used inside coils for transmitters and receivers. The Ferrite core is used to concentrate magnetic lines of force in a peaking coil.

Gain:

The mathematical ratio of how much a signal has been increased or decreased. In electronics, it is the ratio of either voltage, current or power output relative to its input. Mathematically, gain is equal to voltage output over voltage input, current output over current input and power output over power input, etc.

Ground Plane:
A metallic conductive surface used with a vertical antenna. Ground plane increases signals transmitted and received. One half of a signal cycle, transmitted or received, will be directed to or from the vertical antenna, while the other half will go to the metallic conductive surface.

Harmonic:
Any multiples of an original frequency are called harmonics. For example, a fundamental frequency may be one thousand cycles per second of time. The third harmonic of this fundamental frequency is three times one thousand cycles per second of time or three thousand cycles per second.

Heatsink:
A metal, usually finned material, used in mounting electronic devices capable of absorbing heat. The heatsink will transfer (absorb) the heat from the device (such as a transistor or integrated circuit) to sustain it from being destroyed by its generated heat. Without proper heatsinks, heat generating devices will have a much shorter life than those that do have heatsinks.

Impedance:
Opposition to the flow of an alternating current. It is an alternating current resistance in a conductor. The degree of impedance varies with frequency and resistance of the conductor.

Integrated Circuit:
Any combination of resistors, transistors or capacitors in large quantities packaged in plastic, ceramic and/or a metal case. The case will usually have multiple leads attached to it.

Magnetism:
The ability or property to attract steel or iron. Also, current may flow in a conductor, resulting in a magnetic field perpendicular to the conductor.

Microvolt:
Micro refers to one/one millionth. One volt is one unit of electrical pressure. So a microvolt is one/one millionth of one volt or .000001 volts. Micro can also be used in conjunction with current as well as voltage.

Millivolt:
Milli refers to one/one thousandth. Volt is one unit of electrical pressure. So a millivolt is one/one thousandth of one volt of electrical pressure or .001 volts. Milli can also be used in conjunction with current, as well as voltage.

Modulation:
Changing a carrier's voltage at an audio frequency rate (tone, voice, et cetera).

Oscillator:
An electrical device used to establish an electrical wave form. The amount of time this wave form shapes and repeats itself over a unit of time is determined by components used. The repetition of the waveforms over a unit of time is called oscillations.

Oscilloscope:
A "television" type of instrument that gives a visual representation of an electric signal under test.

Phase:
The angular relationship of one waveform to another waveform with respect to time.

Phase Locked Loop (P.L.L.):
A circuit used to maintain a stabilized output of a frequency. It does this by the following method: one signal is compared to a reference signal via a comparator circuit. If one signal does not compare in frequency and phase to the reference frequency, a difference in phase and frequency will cause a voltage change onto a voltage control oscillator. This voltage change will change the frequency of the voltage controlled oscillator and this signal is fed back to the divider and comparator circuit to correct for no phase difference between the reference oscillator frequency and fixed divider output frequency. The loop is locked when no changes in phase is present between the reference oscillator and fixed divider output.

Polarity:
Two opposite charges, one positive and one negative. Also, in magnetics, it refers to the permanent fields of a magnet that are the north and south poles.

Radio Frequency (R.F.):
This is an electromagnetic wave frequency, usually about 30, 000 cycles or greater of an alternating current per second.

Receiver Sensitivity:
The ability to receive a relatively weak signal is a measure of a receiver's sensitivity. When a receiver is turned on, the volume is turned up to a comfortable audio level. When no intelligence is heard, only noise, a signal from a generator is injected to the receiver's antenna jack at a frequency the receiver is tuned to. A signal is applied by means of a calibrated dial on the signal generator (usually calibrated into microvolts) into the receiver. When the receiver goes quiet over the noise when the generator's signal level is increased, this will be the receiver's sensitivity. The signal level is read directly on the dial of the signal generator.

Resistance:
In electrical terms, resistance is the opposition to the flow of current or flow of electrons in a conductor.

Root Mean Square (R.M.S.):
The effective value of an alternating current needed to create the same heating effect of a direct current. The effective value of an alternating current is .707 multiplied by peak value of an alternating peak to peak voltage or current. Mathematically derived, it is the square root of the summed and averaged individually squared sine of each angle of electrical unit (voltage or current being used during one half cycle of time).

Selectivity:
The ability to select a frequency (channel) without interference from adjacent channels and other frequencies.

Sidebands:
Frequency bands above and below the carrier frequency. Lower side bands are carrier frequency minus modulating frequency. Upper side bands are carrier frequency plus modulating frequency.

Sinewave:
A varying and reversing voltage and current that will form both positive and negative peaks of voltage and current over a period of time.

Splitter:
A device used to split and distribute a signal to more than one radio receiver. Internal circuitry is used to overcome losses due to signal distribution to the radio receivers.

Standing Wave Ratio (S.W.R.):
A mathematical ratio of the transmitted radio signal going to a resistive load and what signal is reflected back to the source from the load. This ratio is in terms of voltage, current and impedance and can be measured anywhere on the transmission line at any point in time.

Synthesizer:
The process of combining signals, usually two, that result in one output from either the sum or difference of the signals being combined. The term mixed is often used, rather than combined.

Transformer:
Two coils of insulated wire overlapping each other, having two input leads to one coil and two or more output leads from the other coil. An alternating current to the input of one of the coils will yield an output of the other coil by means of electromagnetic coupling. The output current of the coil windings will be either increased or decreased by means of the number of winds of coiled wire. This will dictate if the output current is "transformed" either up or down relative to the input current.

Transistor:
A device that acts as a variable resistor driven by the changes in electrical current. A transistor is used as a device to amplify voltage or electrical current signals and as an electronic switching device.

Volts:
The amount of electrical pressure to move electrons through a conductor.

Voltage Control Oscillator (V.C.O.):
An oscillator whose frequency can be changed by varying the voltage on a device called a varactor. A varactor is a diode that reacts to voltage changes, changing the diode internal capacitance.

About the Author

Sheldon Chrysler has been a radio technician for over twenty-five years. He received a Bachelor of Science degree in Mechanical Design with a minor in mathematics from Metropolitan State College in Denver, Colorado. He also has an Associate Science degree in Electronics Technology from Denver Community College, Denver, Colorado. While completing his college studies, he earned his commercial Radio Operator's License and Ham Operator's Licence. He has held a Federal Aviation Agency (FAA) Radio repairman's certificate since 1983. Sheldon is credited with designing and creating electronic circuits that are used both commercially and in aviation today. Before opening his own aircraft radio shop in 1982, he worked in several two-ways radio shops servicing C.B. and F.M. land and mobile and public safety radio systems. Currently, Sheldon is teaching adult education in aircraft electronics for the Denver Public Schools, Denver, Colorado.